# Teletext

# Teletext

## Its Promise and Demise

## Leonard R. Graziplene

Lehigh
University
Press

Bethlehem: Lehigh University Press
London: Associated University Presses

Associated University Presses
440 Forsgate Drive
Cranbury, NJ 08512

Associated University Presses
16 Barter Street
London WC1A 2AH, England

Associated University Presses
P.O. Box 338, Port Credit
Mississauga, Ontario
Canada L5G 4L8

The paper used in this publication meets the requirements of the American National Standard for Permanence of Paper for Printed Library Materials Z39.48-1984.

Library of Congress Cataloging-in-Publication Data

Graziplene, Leonard R., 1937–
    Teletext : its promise and demise / Leonard R. Graziplene.
        p.   cm.
    Includes bibliographical references and index.
    ISBN 0-934223-64-5 (alk. paper)
        1. Teletext systems.   I. Title.
    TK6679.7.G73   2000
    384.3'52—dc21                                          00-022427

# Contents

# Preface

IN AUGUST 1981, WHEN IBM INTRODUCED ITS PERSONAL COMPUTER, THE U.S. seemed to be embarking on a new path for its future. In spite of the recession raging in the country at that time, one in which the steel industry in particular was driven to the brink of oblivion, the nation gave signs of being in the last stages of the Industrial Age and in a state of accelerating into the Information Age. Indeed, the PC was only one of several new electro-media technologies beginning to make a run at establishing their own industries. Some of the other most notable of these were videotex, teletext, cellular telephone, and satellite telecommunications.

Most of these did very well and prosper to this day. Teletext, on the other hand, fared the worst of all. Logic dictates it should have survived. Why it failed to do so raises many serious questions. Defeat always invites analysis. Strategic planning, after all, is designed to avoid such negative outcomes. In the case of teletext it appears that either the plan, the technology, or both were flawed. *Teletext: Its Promise and Demise* will address this issue and identify and discuss the conditions that sealed teletext's fate.

The author holds a Ph.D. in Social Psychology from the State University of New York at Buffalo. He is currently a Management Professor at the State University of New York/Buffalo State College. For seven years he produced the CANADA TODAY television series and for twelve years wrote and narrated THE GRAILEN REPORT for radio; a short commentary on the economy and human behavior. In August 1981 he became President of MACROTEL, Inc., a research and development company that specialized in videotex, teletext, and satellite teleconferencing. MACROTEL was a supporter of the Canadian Telidon protocol and became directly involved with broadcast teletext by becoming the page creator for CBS affiliate WIVB-TV in Buffalo when they initiated one of the first local magazines to supplement CBS's national EXTRAVISION service.

# Introduction

IN THE UNITED STATES WE HAVE A LONG HISTORY OF INVENTING, PER-fecting, and utilizing new information technologies. If something useful or better came along, consumers typically bought it in step with raising their standard of living. However, not every innovation successfully navigates the difficult road to consumer acceptance and longevity. It is very difficult to determine ahead of time which new products and services will be successful. For example, in the 1970s CB radios became popular communications devices for our automobiles, but the infatuation proved to be short-lived. During the 1980s almost everyone bought VCRs for their homes, even though our choice of the VHS rather than the Beta format was flawed. And now, at the advent of the new millennium, with liter-ally dozens of new electronic technologies to choose from, it is just as difficult to predict which ones will merit the acceptance of con-sumers and bring profits for their backers.

Often we are surprised to learn that in many information tech-nology areas the United States is not the most advanced, often be-cause of slowness or reluctance to change on the part of vested interests. France, for example, has replaced telephone books with "Minitel" computer terminals. These can bring up recently modi-fied telephone number listings on the system almost instantane-ously. Contrast this with our badly outdated practice of annually printing and distributing telephone books; all of which are already somewhat obsolete the day they are delivered to customers.

While the U.S. remains the world leader in many areas of tech-nology, it is perplexing to understand why some innovations are av-idly accepted and purchased by consumers while others are basically shunned. A case in point is a telecommunications tech-nology that won acceptance in Europe, especially the United King-dom, but which was unable to gain widespread acceptance in the United States. It was called teletext. Several major television net-works and a few cable systems introduced the service to their view-ers during the early 1980s.[1] Teletext should be a key provider of information-on-demand today, but it isn't, because need and prac-

9

ticality were eclipsed by mistakes, broken promises, and high costs. Because of the important lessons it contains for formulating and implementing business strategies, we will closely scrutinize broadcast teletext for clues on how certain internal and environmental factors negatively impacted an entire industry.

In this book we will trace the short twenty-year history of teletext. We will also describe how it was developed, presented to television audiences, and eventually abandoned. An effort has been made to avoid most of the specific technical details pertaining to teletext inasmuch as this book is mainly concerned with being a historical account of the industry and answering the question of why a promising high-tech industry could fail in spite of considerable promise. The book has also been written from a strategic management perspective. Consequently, a great deal of attention focuses on issues and events that impacted the longevity of this industry.

Many people probably feel that an analysis of an old, failed technology has no merit. While it is easy to draw this conclusion, one must be mindful of the fact that throughout this century business has demonstrated numerous consistent patterns. Under similar conditions it is common to find parallel outcomes. We can not only learn from success, but also from failure. In industrial societies, technologies are built one upon the other, with earlier discoveries and accomplishments providing the foundation upon which later improvements can be built. When a new product or service is introduced into the marketplace it is usually promoted with euphoria and enthusiasm. Unfortunately, the success of business is measured in profits, not emotion. In the end, survival and growth are elusive in the absence of commitment to logic and common sense.

What history tells us, however, is the fact that most new technologies fail to make an impact and therefore disappear. When they do, there are always serious questions raised. Perhaps these have transfer value. Maybe failed technologies should become a type of role model for a later, somewhat similar technology. In an effort to avoid a repetition of mistakes, entrepreneurs and managers should never completely dismiss the lessons of the past prematurely.

This could be the case with teletext. For example, there are parallels between teletext and the Internet. The issues faced by both are similar—slow access time, confusing regulatory issues, inconsistent service providers, software incompatibility, unavailability of service at times, and aggressive advertising issues are several of the most evident. In the final analysis it would not be too surprising to see the Internet become embroiled in the same whirlpool of confusion that plagued teletext.

Entering a new industry during its embryonic stage is highly risky. It is safe to say that most new industries fail to survive and for those that do, it is often later entrants, not the earliest participants, that make the most money as the industry matures. While there are never any guarantees that a business effort will succeed, certain precautions must be taken, especially during strategic planning. It is tempting to draw parallels using examples from the past where it appears a particular pattern developed which closely matches current conditions. This tack can be dangerous if taken to extremes. Hardly anything ever replicates itself. Over time the market and the players are always different than their predecessors.

What then should a company do to insure that it maximizes its potential for success in new ventures? There is no simple solution or short list of answers. However, during the process of formulating strategies it would appear essential to at least start from the premise of using good common sense. Certain questions must be raised and honestly answered. There are always obstacles to overcome, and what one wants to avoid is a situation where they are almost impossible to surmount or circumvent. For this reason any analysis of the future must start from the premise of getting a picture of the worst case scenario. If it looks too formidable or the necessary resources are unavailable to make the extra effort to overcome it, perhaps the wise decision is to avoid or delay entry.

In the case of teletext, specific decisions, actions, and both technological and market forces molded teletext's fate. This mix of variables led to the demise of teletext and it is likewise possible that a similar mix of factors could negatively impact present or future fledgling industries. For example, the Internet industry in particular may currently be in a similar set of circumstances that teletext once was in the early 1980s.

Since this book is the account of a promising industry that quickly failed in the United States, it might be a good idea to approach reading the book from the point of view of making it a learning and planning exercise. This would be especially useful for individuals who are, or will be, involved in new business ventures. To assist the reader in this task, seven strategic areas have been identified which have relevance as teletext evolved through its brief life cycle. These are listed below.

SEVEN RELEVANT POINTS TO CONSIDER:

1. Determine if there is a real need for the product/service.

2. Gauge the magnitude of anticipated demand for the product/service.

3. Decide how to reach and stimulate growth of the market.

4. Be sure laws are conducive for nurturing the growth of the product/service.

5. Determine if needed materials are available in sufficient quantities and at reasonable prices.

6. Attempt to identify other participants with whom an alliance can be formed to minimize financial risk.

7. Always believe half of what you hear. The half you don't listen to is made up of exaggerations and outright lies!

# Teletext

# 1

# Teletext: A New Idea for News Dissemination

DURING THE EARLY 1980S THERE WAS A CONSIDERABLE AMOUNT OF speculation on the part of information dissemination companies and industries over who would dominate the homes of the future market. Among the strongest of those identified by the Yankee Group, a highly respected marketing research firm from Boston, were AT&T, cable television, the electronics industry, the publishing industry, and broadcasting. They drew a comparison among these leading contenders with regard to perceived strength in eleven key areas (see Table 1.1).

According to this report, as of August 1982 the company with the number one current market position was AT&T, followed by Sears and the cable industry. What is particularly interesting about the rankings is the fact that they represent no less than seven different industries. These include the telephone, electronics, publishing, cable, broadcasting, retailing, and financial services industries. All, however, had staked out their own claims to becoming the dominant information source in the home of the future. Little did anyone realize at the time that shortly all of them would be involved in an attempt to control a technology that gave promise of becoming the elusive key to dominating the home of the future. That technology was teletext.[1]

## NEWS ON DEMAND

How would you like to be able to turn on your television set at any time of the day or night and in seconds select and watch your choice of the latest updated news, weather, or sports? No, we are not talking about CNN or Headline News. These are marvelous services, but they, not the viewer, control the content of what is seen on the screen at any given moment. What we are talking about is a service that permits viewers to see, within seconds, their choice of

15

**STRENGTHS AND WEAKNESSES OF LEADING CONTENDERS IN THE HOME OF THE FUTURE MARKET**

| Position of Strength | AT&T | Cable Industry | Sony | Tandy | IBM | Dow Jones | Knight Ridder | CBS | Amer. Expr. | Sears | GE |
|---|---|---|---|---|---|---|---|---|---|---|---|
| 1. National Distribution | | | | | | | | | | | |
| 2. Service and Maintenance | | | | | | | | | | | |
| 3. Entertainment Software | | | | | | | | | | | |
| 4. Information Bases | | | | | | | | | | | |
| 5. Financial Services | | | | | | | | | | | |
| 6. Technology | | | | | | | | | | | |
| 7. Direct Electronic Home Access | | | | | | | | | | | |
| 8. Hardware Manufacturing | | | | | | | | | | | |
| 9. Capital Resources | High | High | Low | Low | High | Low | Low | Med | High | High | High |
| 10. Willingness to Compete | High | High | Med. | Med | Low | Low | Low | Med | High | High | High |
| 11. Importance of Home Strategically | High | High | High | High | Med.-H | Low | Med | High | Med | High | High |
| 12. Present Market Position | 1 | 3 | 7 | 4 | 10 | 9 | 7 | 5 | 6 | 2 | 8 |

**Table 1.1. Home of the Future Assessments. The Yankee Group—Boston.**

reports. Sounds good, doesn't it? You may be surprised to learn that during the mid-1980s we were very close to having such a national service, virtually free of charge.

The technology was called teletext. Simply stated, it involved using a portion of the broadcast signal called the vertical blanking interval (VBI). Nontechnical consumers will recognize its presence as the bar separating identical scenes when the picture on the TV screen rolls because of an improperly tuned vertical control. For those more technically oriented, the teletext signal was contained in lines fourteen through seventeen and twenty of the twenty-one lines of the 525 used in NTSC video (line twenty-one is used for closed captioning for the deaf).[2]

Basically, what teletext did was make it possible to transmit digital text and graphics in color simultaneously with normal television programming. There were basically two ways to access teletext. One was through connection to a CATV station. The other was by accessing a VHF television signal. In the latter, no cable was required; the signal was transmitted through the air. Once you turned on your television set you had the choice of selecting either regular video programs or teletext.

In the early 1980s three North American television networks provided broadcast teletext services; CBS, NBC, and the CBC in Canada. In addition, several other smaller broadcasters, mainly PBS, were involved in providing a teletext service. All that was needed to receive the service was a television set, a decoder, and a remote keypad.[3]

If one had this equipment and teletext was being carried by a local television station, it was relatively easy to access teletext (one could actually see the transmitted data as small pulsating lines in the vertical blanking interval). The viewer simply pushed a button on a remote keypad to switch from the network or locally originated video program to the main menu screen of the teletext service. The configuration used in CBS's EXTRAVISION service, as seen in Figure 1.1, was typical. Once accessed, a main menu appeared and all the viewer had to do was push the appropriate numbers and/or buttons on the keypad to receive the specific news item desired. Then, after reading the accessed teletext pages, pressing a button on the keypad would quickly return the viewer to the video program. In Figure 1.2 the left column represents the teletext page one might access by cutting away from the regular television programming seen in the right column. The one exception is the closed-captioning template seen in the left column—second from the top.

Reading
Between The
Lines

A viewer uses her
EXTRAVISION™ decoder to
turn to any of 100 pages of
national and local
information available 24
hours a day.

Figure 1.1. Accessing **EXTRAVISION. CBS—New York.**

**Figure 1.2. EXTRAVISION Pages. CBS—New York.**

Each television network that originally provided the service had staffs regularly updating national and international news stories throughout most of the day. In addition, a few network affiliates provided local news magazines as an added service. To stimulate interest these stations also placed public access kiosks in shopping malls in their viewing area to acquaint consumers with the service.

## The Characteristics of Teletext

What viewers who had access to teletext saw on their television screens was quite different than any other form of news dissemination ever before previously available.

For one thing it was a silent medium. There were no narrators. Stories had to be read from the television screen. This was just the opposite of how television is perceived to influence viewers. In some respects it was like reading a cross between cartoon caricatures and print. A novel effect was created by the judicious use of different colors and accompanying graphics carefully blended to accompany stories.

This type of presentation was very strange because teletext looked more like what is normally seen on a computer screen rather than television. Indeed, from a technical perspective, this is exactly what was happening, because as one made the switch from television to teletext, the modes of delivery changed from analog to digital. If one didn't mind the reading and silence associated with teletext it more than made up in content and convenience for what it lacked in familiarity.

Teletext was essentially a one-way service. Viewers could access information of their choice, but could not return messages of their own to the service provider. On the other hand, videotex, a service comprised of basically the same protocol as teletext, was a two-way service. With the assistance of a modem and telephone line, it permitted financial transactions to occur. It was therefore well-suited to enable banking and shopping at home. While it was confusing for some to make the distinction between the two technologies, there was a distinct difference in the way each worked. These differences are diagrammed and compared in Figure 1.3.

During the developmental stages of teletext in the U.S. it was widely felt that because of AT&T's being broken up by the federal government, telephone rates were going to go up substantially. If that ended up being a reality, then broadcast teletext would become more attractive than videotex. According to Richard Neus-

▼ *A representative Videotex system.*  ▼ *Representative Teletext and Cabletext systems*

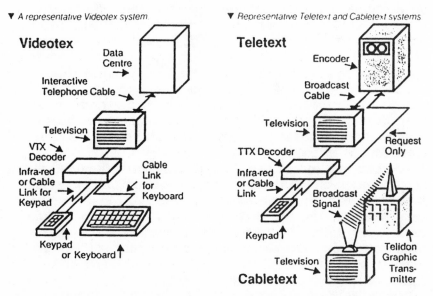

Figure 1.3. **Videotex and Teletext System Configurations. Canadian Department of External Affairs—Ottawa.**

tadt, a Washington lawyer and former communications adviser in the Carter administration: "As the copper wire costs go up, the over-the-air costs should come down. . . . Thus, being positioned in the broadcast teletext market would place you on the right side of the cost curve."[4]

The fact that teletext itself was free, especially during a time when costs for all forms of information were escalating, made it seem all the more certain that it would quickly become a favorite with television viewers. Unfortunately, teletext was a bust. All within the span of a few short years, the national teletext services of the CBS, NBC, and CBC television networks passed from the scene uneventfully. So too did the cabletext services of *Time* and KEYFAX as well as most of the lesser ambitious teletext services offered by PBS and smaller cable enterprises.

In order to understand what happened to teletext it is necessary to trace its birth and evolution. As you will see, right from the beginning there were signs of trouble.

## THE EARLY DAYS OF DEVELOPMENT

Teletext was the name given to systems that distribute text information on spare transmission lines in broadcast television signals

for decoding and presentation on specially modified television receivers. The basic idea originated around the mid-1960s when RCA
developed a system called HOMEFAX. It used a spare line in the
field blanking interval of the television signal to transmit a slow-
scan image from a camera, which was then read as hard copy at
the destination site. A whole frame, (one full static page of information on a TV screen), of information took about ten seconds to deliver. By the early 1970s, HOMEFAX was discontinued because of
improvement in the technology and the absence of a sizable demand for the service.

The BBC and Independent Broadcasting Authority (IBA) in the
U.K. picked up on the idea and started developing fledgling teletext
services. Their approach improved on HOMEFAX in two major
areas. One, the text was digitally encoded so that pages could be
read directly from a computer, which meant that a whole row of
text information could be carried on a single line of the broadcast
signal. Two, text was presented directly into the receiving television
screen rather than as a hard copy facsimile.

Initially, teletext was used by the two services for internal communication, but in 1974 the BBC started test transmissions for a
public audience. They called their service CEEFAX (the name de-
rives from "See Facts"). A year later, IBA started test transmissions
of its service, which it called ORACLE (Optical Reception of Ac-
couchements by Coded Line Electronics). Both names accurately
defined the objective of teletext.[5]

At first there were no commercial teletext decoders available for
a test audience to view the services. Therefore, the earliest users
were home electronics enthusiasts who had built their own decoder units as add-ons to their television receivers using plans published in a popular electronics and wireless hobbyists' magazine.
Several independent companies began to manufacture decoder
units based on the same plans, but only on a small scale. This nar-
row interest negatively impacted the further development of teletext for several years. The services failed to grow because of the
unavailability of reasonably priced decoders. It wasn't until 1978
that the stalemate began to dissipate.

Since most homes in the U.K. rented rather than bought television sets at that time, it was the rental companies that began to
push the availability of teletext decoders as an added revenue
stream. Under this arrangement, by 1979 teletext audiences grew
swiftly. In one survey, general awareness of CEEFAX grew from 10
percent to 45 percent of television viewers. By one estimate, by
mid-1979 there were twenty thousand teletext receivers in the U.K.

and one hundred thousand a year later. In spite of this rather phenomenal growth, the latter figure represented only one-half of one percent of all households in the United Kingdom with television sets.[6] Table 1.2 contains a listing of the various receivers available in 1981.

## SORTING THROUGH THE DIFFERENCES

A great number of companies that became involved with teletext in the U.S. were prominent leaders in the communications industry. Several of the most notable of these were CBS, NBC, *Time* and Zenith. In addition, there were also KEYFAX, a joint venture involving Satellite Syndicated Systems, Keycom Electronic Publishing, and Ameritext, the U.S. marketer of the World System teletext standard. On this last point, it is important to identify the two incompatible standards which were attempting to become the favored protocol in North America.

The first, and older, system was the World System developed by the British Broadcasting Corporation. The second was the North American Broadcast Teletext Standard which basically evolved from the Canadian Telidon system. Given the fact that the number of scan lines in British television was 625 and in North America it was 525, it is not surprising that there would be further differences between the two approaches to television-based services. The World System was synchronized with the lines of the horizontally transmitted TV signal. NABTS, on the other hand, was transmitted asynchronously and required greater error correction. The World System was less expensive, but NABTS had superior graphics quality.[7]

In the early 1980s, systems using both standards began to proliferate in North America. Each apparently felt their approach had the most merit, but because of their technical differences, it appeared inevitable that there would have to be agreement and consolidation or else teletext would have difficulty succeeding. Many alliances were formed and these are discussed in later chapters.[8]

Because teletext originated and achieved its earliest success in Great Britain, a comprehensive and primary analysis of the technology must start there.

| MANU-FACTURER | CONTACT—NAME, ADDRESS, PHONE No. | | MODEL No. | SCREEN SIZE | PRICE |
|---|---|---|---|---|---|
| DER | Marketing Dept. DER Apex House, Twickenham Road, Feltham, Middlesex, TW13 6JQ | (01) 894-5555 | 6213 6053 5283 | 16" 20" 22" | £10.95 per £12.45 Month £13.95 Rental |
| GRANADA | Marketing Dept. Granada TV Rental, P.O. Box 31, Ampthill Road, Bedford | (0234) 55233 | 16BZ4 22BZ7 26BZ7 | 16" 22" 26" | £10.95 per £15.95 Month £17.95 Rental |
| GRUNDIG | Paul Agate, Grundig International Ltd Mill Road, Rugby, Warwickshire, CV2 11PR . | (0788) 77155 | T511401 T561401 T661401 | 20" 22" 26" | £289.95 inc. £329.95 VAT £439.95 |
| | (ADAPTOR BOARD VT100fKGBc. £60.00 IS NECESSARY FOR ALL MODELS) | | | | |
| HITACHI | Consumer Unit, Hitachi Sales (UK) Ltd. Station Road, Hayes, Middlesex. | (01) 848 8787 | CPT 2060 CPT 2260 CPT 2660 | 20" 22" 26" | £422.17 inc. £466.61 VAT £522.17 |
| JVC | Terry Atkins, JVC (UK) Ltd. JVC House, 12, Priestly Way, Eldonwall Trading Estate, Staples Corner, London NW2 7BA | (01) 450 3282 | 7742T 7842T 7943TS | 20" 22" 26" | £399.00 £439.00 inc. £609.00 VAT |
| MITSUBISHI | Sales Dept., Mitsubishi Electric (UK) Ltd., Hertford Place, Denham Way, Maple Cross, Rickmansworth, Herts WD3 2BJ | (0923) 770000 | CT2101TX CT2230TX CT2627TX | FST 51cm 22" 26" | £445-495 £419-471 inc. £525-589 VAT |
| NATIONAL PANASONIC | Sales Promotion Dept. 300-318 Bath Road, Slough SL1 6JB | (0753) 34522 | TX1642 TX3300 TX2646 | 16" 22" 26" | Varies according to Dealer |
| PHILIPS ELECTRONICS | Clive Parkman 420-430, London Road, Croydon, Surrey CRG 3QR | (01) 689 2166 | 4616 3745 3890 | 16" 22" 26" | Varies according to Dealer |
| PYE LTD. | Sales Dept. St. Andrews Road, Cambridge, CB4 1DM | (0223) 354262 | 4146 5447 3757 | 16" 22" 26" | £339 inc. £389 VAT £499 |
| RADIO RENTALS CONTRACTS | Radio Rentals Contracts Ltd. Apex House, Twickenham Road, Feltham, Middlesex, TW13 6JQ | (01) 894 5644 | 6023 8233 6103 | 20" 22" 26" | £292.00 £343.50 + VAT £426.50 |
| REDIFFUSION | Clair Daly Doric Radio, Fullers Way South, Chessington, Surrey | (01) 397 5411 | DORIC CU51406D CU56406D CU67406D MURPHY 2003TE 2203TE 2603TE | 20" 22" 26" 20" 22" 26" | See Individual Dealer |
| ROBERTS RADIO | Jean Smillie, Roberts Radio Co. Ltd., Molesey Avenue, West Molesey, Surrey KT8 0RL | (01) 979 7474 | DYNATRON D650TXT D800TXT | 22" 26" | See Individual Dealer |
| SONY | Customer Services, Sony House, South Street, Staines, Middlesex, TW18 4PF | (01) 61688 | KV 2066 UB KV 2256 UB KV 2766 UB | 20" 22" 27" | £504.65 inc. £712.80 VAT £734.40 |
| TATUNG | Publicity Dept. Tatung (UK) Ltd., Stafford Park, 10, Telford, Shropshire TF3 3AB | (0952) 613111 | TP9404 TT9409 TV9410 | 16" 20" 22" | Models in process of replacing/up-dating call for details |
| THORN EMI FERGUSSON | Ann Waterman, Thorn EMI Fergusson Ltd. Cambridge House, Great Cambridge Rd, Enfield, Middlesex EN1 1UL | (01) 363 5353 | 37023 2263 37373 | 16" 22" 26" | £299 inc. £399 VAT £469 |
| TOSHIBA | Jacqueling King, Toshiba (UK) Ltd, Toshiba House, Frimley Road, Frimley, Camberley, Surrey, GU16 5JJ | (0276) 62222 | 22 IT 26 IT | 22" 26" | £399.95 inc. £559.95 VAT |
| ZANUSSI | Jim Woodgate 82, Eaversham Road, Zanussi House, Reading, Berks, RG1 8DA | (0734) 470011 | 22ZA377 22ZA577 | 22" 22" 16" 26" | £399.95 £499.95 need Adaptor Board, KTV 665. £59.95 inc. VAT |

Table 1.2. British Teletext Receivers.

# 2

## Teletext Developments in Europe

### THE BRITISH TELETEXT STANDARD

TELETEXT WAS CLEARLY A BRITISH INNOVATION. ACCORDING TO AN IN-dependent survey sponsored by the Information Services Division of British Telecommunications, there were 774,385 British standard viewdata and teletext sets in operation in fifteen countries around the world. This accounted for 98 percent of all sets in the world at that time. Although viewdata was a two-way service, somewhat different than teletext, the point worth making here is the fact that the British standard pretty much had a monopoly. Chip sets from Britain were being distributed in six different languages in 1980: European English, U.S. English, French, German, Italian, and Swedish/Finnish.[1]

To further strengthen the position of the British standard, its inherent qualities seemed to offer everything necessary to satisfy all users of teletext. For example, in the British standard:

A. Viewdata and teletext were fully compatible with each other because a combined chip set allowed the same display screen to access broadcast teletext and viewdata.

B. It was designed for reception on both 625 line 50 field TV sets and 525 line 60 field TV sets.

C. The application level was alphamosaic, but upgradable to other alphabets and graphic characters.

D. The standard allowed for seven colors in addition to black and white.

E. The standard was designed to provide low-cost displays for the consumer market.

F.  The standard was page-oriented, that is to say, it displayed one page at a time.

G.  A page of information consisted of a maximum of 960 characters in 24 rows of 40 characters, which translated to 75 to 100 words per page.

H.  It was easy to use.

I.  The British standard was designed to be fully upwardly compatible from the then-current Level 1 to the future commercial Level 5.[2]

Perhaps we should elaborate on this last characteristic because it was to play such a prominent role in the future acceptance of the British standard, particularly in North America. To do so, a brief description of the various levels of teletext will reveal some major differences and establish the range of possibilities for the appearance of teletext:

Level One had alphamosaic alphabet and graphic characters.

Level Two introduced one full character set for all Roman-based alphabets.

Level Three introduced Dynamically Redefinable Character Sets (DRCS) which allowed high-definition graphics.

Level Four allowed for an alpha-geometric display.

Level Five presented the ability to store, distribute, and display full color still pictures using data compression techniques developed for slow-scan teletext.

The British called their Level Five configuration "Picture Prestel." While Prestel was a two-way videotex service, somewhat distinct from teletext, the viewdata and teletext standards were compatible. Consequently, teletext too was positioned to present slow-scan pages. However, the problem for broadcast teletext was bandwidth. Transmitting slow-scan pictures in the early eighties required hard wiring. In other words, ISDN, fiber-optic cable, or cable TV networks were needed for sufficient bandwidth.[3]

While the British might have been justified for avoiding Level Five teletext, there is some question why they stopped short of Level Four. Where the British standard ran into difficulty from the beginning was their adamancy in concluding that there was the danger of too much concentration on display standards, and not enough concentration on other equally important standards such as user and gateway protocols. In other words, the British were more preoccupied by the need for uniform user commands for menu and keyword searches to help the consumer. Equally important, to facilitate access to more data services cost-effectively, gateway protocol standards had to be addressed.[4] What the British didn't realize was the vast importance the rest of the world placed on display standards. In the end this difference fragmented the teletext industry and created a great deal of avoidable confusion.

## BRITISH RESEARCH QUESTIONS

Teletext appeared to have a promising future in the late 1970s. However, in spite of the optimism that teletext was destined to become a major news disseminator and a mass market service, more research was still needed to learn additional facts about viewer habits and satisfaction.

One of the most influential studies of teletext took place in 1980. It was conducted by Communications Studies and Planning Ltd. and Eastel, the major Prestel Information Provider in the U.K. A sample of 369 teletext households was drawn and 268 of these completed diaries, recording details of their household's use of teletext over a one-week period in September 1980. Later, 245 teletext users participated in a more thorough interview analysis in which they were asked about the acquisition process, the patterns of household usage of teletext, and perceptions and criticisms of teletext.

The purpose of this research was to assess the extent to which electronic information services served a useful purpose in the residential market and to examine the extent to which improvements to service features were needed. More specifically, the key questions to be answered centered around the following:

1. Was there a clear audience for electronic publishing?
   —Who buys or rents teletext sets, and why?
   —Who uses teletext in the household?
   —What is it used for?
   —When is it used?
   —Is teletext just a cheaper version of videotex?

2. What service characteristics were of key importance?
   —How useful are graphics in teletext presentations?
   —What is the optimum trade-off between database size and average access time per frame?
   —How acceptable is the text legibility and what problems are experienced in using teletext?
   —Are additional facilities and capabilities of real benefit?
   —What other capabilities would users like to see on their teletext sets?

3. What were the broader implications?
   —How does teletext affect general television viewing?
   —Does the selection of CEEFAX or ORACLE depend on previous television channel viewing and/or subsequent channel choice?
   —Is teletext used as an alternative to the commercial break?
   —Does the use of teletext influence use of printed media such as books and newspapers?[5]

## SUMMARY OF BRITISH RESEARCH FINDINGS AND CONCLUSIONS

Based on the data of this survey, teletext was said to have a good chance of success, particularly among professional and managerial types, perhaps because at the time they accounted for most of teletext's customers. Teletext users accessed the service on average more than once a day and were generally pleased with the information provided. Levels of satisfaction and perceived monetary value were high. On the downside, viewers would have liked to have seen an increase in the speed of frame access and a larger database of information.

Some of the more significant specific findings of the study were as follows:

A. Teletext users were primarily male adults between the ages of thirty and fifty in managerial and professional occupations.

B. Teletext users appeared to have a higher than average need for or interest in information.

C. Higher than average interest in information was accompanied by an unexpectedly high level of usage, in teletext owning homes, of other novel forms of electronic equipment.

D. Ninety-two percent of respondents said their teletext sets were reliable.

E. Teletext services were most often used for news, weather, travel, and sports news. On average, a teletext service was accessed 1.4 times per day and the mean number of pages used on each occasion was 4.7.

F. Accessing financial information had the highest number of pages per occasion—2.4.

G. Peak use of teletext was around 6 P.M. There was a fairly consistent level of use day by day in terms of accesses, but during the week accesses were less often for sports results than for news and other data. On the weekend, the pattern was reversed.

H. About 10 percent of teletext access occasions were less than two minutes long, while 48 percent lasted from two to ten minutes. Only eleven percent lasted longer than thirty minutes. Most teletext accesses were, therefore, long enough to disturb normal TV viewing since they could not be fitted into a typical commercial break of about two minutes.

I. Ninety-three percent of users said they were very satisfied or satisfied with teletext and 85 percent said it was a very good or quite good value for the money. On the other hand, 29 percent described teletext as incomplete; 25 percent as too general.

J. The most commonly identified area in need of improvement was a faster page/frame delivery, named by 77 percent of respondents. The need for fuller illustrations was cited by 65 percent.

K. In contrast, there was relatively little interest in a reduction of the service. And, only nine percent of users were interested in a smaller screen size than the 22 to 26-inch typical teletext set. It should be pointed out that the desire for information was not, in most cases, the primary reason for acquisition.

L. Overall, there was almost universal support of teletext as evidenced by the high level of probable renewal of rental contracts (84 percent expected to renew their rental contracts).[6]

The study pointed out the fact that many of the users interviewed were innovators—among the first to acquire new products and services—and were not typical of a later and more widely diversified market. Nevertheless, the data from the interviews and diaries suggested that teletext had a good chance of becoming successful.

## SORTING THROUGH PROBLEMS

As for the market itself, there were both good and bad factors to deal with. On the plus side, it was reported that the whole annual budget for a nationwide, two-network teletext service was less than the cost of a single television program or major drama. Spelled out differently, the actual cost to each of the two services for broadcasting teletext eighteen hours a day, 365 days a year, was about $500,000 per year. For their money, the information providers reached half a million homes that averaged two hours of viewing per week.

On the downside, there was still the matter of very slow page access. If two lines were being used to pass one hundred pages of teletext information, it would take from one-quarter of a second up to thirty seconds for the requested page to show up on the screen. When four lines were used these times were cut in half; from one-eighth of a second to fifteen seconds. And, if eight lines were utilized, a number authorized by the government, access times would again be cut in half to between one-sixteenth of a second to 7.5 seconds. It was felt these latter times were acceptable to the majority of viewers. However, in the early 1980s the information providers were using only four lines and viewer patience was being strained. To compensate, some viewers would request the next page immediately after their previously requested page appeared on the screen; using the lag time before the second page appeared to read the first page. Of course, this did not always work satisfactorily, especially when a second page came up, for example, a second after being called for.[7]

There were those who criticized teletext because it wasn't interactive; a quality believed to be highly desirable. Not to be stymied, some prominent British executives associated with teletext tried to make the case that broadcast teletext was more interactive than it seemed.[8] If it were not enough to call attention away from a negative feature, still other executives made the claim that teletext was converging with videotex; the two-way cousin of teletext.[9] Neither carried much weight and as anyone who used teletext would attest, these claims were more a marketing ploy than a statement of fact. However, it was yet another aggressive attempt to solidify teletext's rather weak but promising status in the minds of British consumers.

## SEARCHING FOR VALUABLE USES

Teletext was first and foremost a business, but it was also a medium in search of an audience that would be rewarded from its use. That is why some futurists began to define teletext in terms of its most valuable different uses. Some called it a great hobbies medium; one that not only could serve deaf special interest groups, but also one that could do the same for anglers, artists, florists, or dozens of other hobbyist groups.[10]

Considerable numbers of optimists in the U.K. felt that there was money to be made from teletext. What inspired this belief was the speculation that by 1990 there would be ten million teletext sets in place.[11] Making money with teletext was a major challenge from the beginning in Britain. For one thing, advertising was restricted by the government to 15 percent of the total pages available. What helped teletext develop and grow in Great Britain was the fact that the majority of teletext sets were rented. Since the rental price for a teletext equipped set was only around a pound more per week than a regular TV set, the cost to consumers was minimal when compared with the alternative of having to purchase a decoder to view teletext.[12] While this incentive helped increase the number of teletext viewers, advertising remained a prominent part of the strategic plan in spite of the limitations placed on its role.

A significant advantage teletext had over the print media was its up-to-the-minute news information. Still another lucrative enticement was the widespread consensus that teletext would be an advertising gold mine. And finally, the usefulness of teletext to retailers and banks to promote special offers capped off the wave of enthusiasm. Of course, unless there is sufficient cash flow no new industry can survive for very long. While it is understood that in the earliest stages of industry development there is seldom enough to make very many participating companies a profit,sooner or later something has to click and customer spending must escalate.

Another possible anticipated valuable use for teletext in Britain was for supermarkets to promote special offers to shoppers. One marketer suggested that advertisers hide clues in their teletext commercials and make the public work for the answer; offering prizes of their product as a reward. Still another company spoke highly of using teletext as a medium for downloading computer software from teletext directly into a home computer; a process that became known as telesoftware. Banks too, it was felt, could use the medium for similar purposes. Teletext was also expected to tap the potential of the small local advertiser. The list goes on to include classified ads; birthdays and anniversaries; airport flight times; hotel magazines, and anything and everything that might satisfy the great appetite for information perceived to exist.[13]

## APPEALING TO INTERESTS

A number of tests involved teletext and education in England. In the CEEFAX and ORACLE 19 Schools Project in 1979, special pages were provided on teletext to supplement a number of science oriented TV programs already in use in the classroom. These were available before and after the programs and served the dual functions of notes for the teachers and reminders for the students. Because it was found that teletext increased classroom discussion, the test was later repeated for six junior schools. While it temporarily added a new dimension to teaching and learning, teletext did not become a major tool in education. Indeed, teletext only provoked more technological alternatives that have always been a quandary for education.[14]

However, the preponderance of speculation centered around teletext as an entertainment medium. It was repeatedly pointed out that whatever else people looked for from their television set, they expected it to entertain them. And who was to argue that television did not consistently mean entertainment. But how was one to entertain through teletext? Some of the types of entertainment tried included bingo and treasure hunts for children, but far and away the most successful venture was a Guess-The-Price-Of-Gold competition. Viewers of the financial pages were asked to guess the price of gold in three month's time. The nearest correct answer would win a Krugerrand with a half-Krugerrand for the runner-up and smaller pieces for a few less accurate answers. The first time this program was run 3,555 entries were submitted. Based on the then-stated price of gold, management felt this was a very cost-effective way to stimulate interest in teletext and at the same time acquire the names and addresses of over three thousand teletext set owners who had an interest in gold.[15]

## ORACLE's Plan

ORACLE set four objectives for themselves. In the first place they felt they had to educate the marketplace about their product. Second, they wanted to boost their circulation by raising the number of sets wired to receive teletext. Thirdly, they wanted to gain more viewers as a means of attaining their fourth objective, attracting advertisers.[16]

By mid-1985 the British ORACLE teletext service had already reached a very important clientele in large numbers. Their service, at that time, was being carried into 2.4 million homes. It was claimed that ORACLE reached more college educated adults than the *Financial Times* or the *Times*.[17] Among the features most accessed by businesspeople were foreign exchange rates, stock share price movements, and commodity trading news. Other popular features were company reports and news such as shareholder announcements, interim results, and year-end statements. And, not to stray too far from the broader consumer market, they even had a family finance section.[18] While ORACLE did a fine job of making financial information available, they were later to learn that consumers were much more interested in being able to conduct transactions.

The people running ORACLE felt their service was the nearest thing in existence to an electronic publishing company. That is why advertisers who spent much larger sums of money on traditional publications may have missed an opportunity, particularly if as claimed, businesspersons were more inclined to access teletext.

Immediacy was, of course, a major advantage of teletext over print. ORACLE updated its pages around five thousand times a day or, on average, once every fifteen minutes per page. Furthermore, the service was continuously available twenty-four hours a day. This was information immediacy at its best.[19]

Because up-to-the-minute news could easily be disseminated through teletext, this gave the medium a big advantage over the press. Apparently this was a point well noted by several major advertisers such as Esso, Crosse & Blackwell, Nestle, Sony, and most TV rental companies. This was encouraging, especially when it was estimated that by the end of the 1980s there would be one hundred million teletext sets in Britain which would translate to a yearly figure of ninety million dollars in advertising revenue. Broken down, this would have amounted to a very modest advertising income of nine dollars for each teletext set.[20]

## British Business Teletext

There was interest in teletext on the part of the business community, but the question was repeatedly asked how service providers might generate a revenue stream from this group. Advertising made sense because it was the traditional way money was always raised with television, which of course is where teletext resided. Beyond this option, however, the only other possibility would have been through closed-user groups which had their own specialized interests and needs. Unfortunately, this would have fragmented the teletext signal and allowed some to use the service to the exclusion of others. Encrypting some pages of information might have alleviated the problem, but the weakness of this technology coupled with a limited number of pages that teletext could broadcast at any one time led to closed-user groups having less appeal.

Then too, when it came to accessing business data and information, other technologies were already able to do the same thing. The real competition was not between teletext and print when it came to business, but between teletext and other electronic dissemination equipment. This had to badly constrain the continued growth of teletext for business purposes.[21]


## Developments in France

The major competitor for the British in Europe were the French. They developed their own protocol and, like the British, set out to sell their version to the world. They called their protocol Antiope and as we will see later, the French were quick to establish a presence alongside the British in North America.[22]

Antiope technologies were designed to permit operation not only in teletext and viewdata modes, but in a hybrid mode as well. By 1980 the French were offering full-scale services on a commercial basis for applications such as stock market information, news and weather, traffic reports, and agricultural reports. Other services were added in the next two years and extensive tests involving two hundred thousand terminals were conducted by Velizy and Ile et Vilane (the former a suburb of Paris, the latter a suburb of Rennes). These tests involved Antiope teletext and the first phase of an electronic telephone directory project, which was conceived to provide a terminal (electronic phone book) for every private phone in France by 1990. This number was estimated to be thirty-four million terminals and to assure achievement of this goal, was subsidized by the government.

What made the French somewhat unique is the fact that they were the first to use entire television channels for broadcast videotex. Nowhere else in the world had more than the vertical blanking interval been regularly used by 1980. They were much more domestic in their ambitions than the British, but as early as 1979 plans had been formulated to enter North American markets with Antiope. This was not surprising inasmuch as the U.S. had not developed a teletext protocol (aside from AT&T's prototype) of its own. Canada, of course, had Telidon. Four of France's early successes in the U.S. included:

1. The use of Antiope by CBS in their teletext tests.

2. The use of Antiope by KCET in Los Angeles, the second largest PBS station, to test teletext on UHF channels and to develop educational uses for teletext.

3. The evaluation of Antiope by WGBH, the PBS station in Boston, as a means of providing a new, advanced captioning system.

4. Control Data Corporation's experimentation with Antiope as a means of enhancing its PLATO education computer system.

In October 1979, France and Canada signed an agreement which ran for three years and provided for an exchange of information regarding the Antiope and Telidon systems.[23] Although technically competitors, it was this kind of spirit of cooperation that made one think the industry could grow in an unselfish and supportive environment. In the end France and Canada were key contributors to the NABTS Standard, but the British maintained their own standard and kept the industry highly competitive.

## OTHER DEVELOPMENTS IN EUROPE

As pointed out earlier, the British viewdata/teletext protocol became the dominant standard in the world, and by 1980 a number of other countries in Europe acquired their software and moved ahead with their own plans to offer teletext services. Some of the countries that made an early commitment to the technology included West Germany, Denmark, Belgium, Sweden, and even

Australia. For example, in West Germany they called it Bildschirm-zeitung, and ARD/ZDF broadcasting announced plans for a test service. Beginning in 1981 a considerable number of trials were well underway and it looked very promising for the British teletext standard to continue its domination of the industry.[24]

One of the most interesting developments in teletext occurred in Sweden. The Swedish Broadcasting Corporation initiated trials in late 1979, and by the summer of 1981 there were already in excess of thirty-five thousand broadcast teletext receivers installed in Swedish households. In the opinion of industry leaders in that country, broadcast teletext would in due course be a standard feature in TV sets of certain size classes. Furthermore, they also felt that most sets sold in 1985 would be able to receive broadcast teletext. At the time, the cost of a twenty-six-inch teletext receiver was some two hundred dollars more than the same set without teletext.[25]

Although Sweden was perhaps the most socialistic country in Europe at the time, the fact that they prohibited advertising on teletext because it threatened the earnings of the daily press was still somewhat of a surprise. It was then left to the government to underwrite the expenses associated with providing teletext services.[26]

Iron Curtain countries such as Hungary had teletext trials, but their efforts were several years behind the leading Western countries. In Hungary, the first teletext test transmission went on the air in 1980 and working models of viewdata were operational in 1982.[27]

As of 1 January 1982, there were 175,000 teletext-receiving TV sets in Holland, 110,000 in Austria, 80,000 in Sweden, 150,000 in West Germany, and many more thousands throughout the rest of Europe.[28] This growth was very encouraging for the industry, and if one factored in countries like Australia and New Zealand it soon became clear that teletext was now a global business. Also, at the same time Japan was well advanced in developing their own protocol which could function with their own unique character-set alphabet.[29]

One could go on to describe the evolution of teletext in just about every advanced country of the world. While this would be interesting, it would also be very time-consuming and somewhat repetitious. What is more important is to point out the fact that teletext had a broad-based appeal in many parts of Europe and generated a considerable amount of interest elsewhere. What is surprising is that the United States was not one of the main architects of teletext. In this respect Canada far overshadowed the U.S. and indeed

developed a protocol that cut heavily into British market domi-
nance, in spite of its early lead and adoption by many countries.

It was inevitable that the largest consumer market in the world
would soon become the focus of the British, French, and Canadians. For a fuller description of the features and marketing
approach taken for the British World Teletext Standard, see Appendix 1.

# 3

# Developments in North America

As might be expected, teletext quickly made its way from Europe to North America. The technology made rapid strides in Canada where, like in Europe, the government took a direct hand in promoting and financially supporting teletext. While Britain was the first country to make a viable service of teletext, it was Canada that improved the technology to make it more lucrative to a wider range of audiences. Through the research and development efforts of the Canadian Department of Communications, the technology took on a new character and generated widespread interest. They called their version of teletext TELIDON.[1]

## Canada's Telidon Teletext

Telidon was designed to provide high quality graphics. The secret of Telidon was its coding scheme—the Picture Description Instructions (PDIs). PDIs described the content of images in terms of basic geometric elements such as points, straight lines, arcs, rectangles, and polygons. For these reasons, Telidon was described as alpha-geometric. Telidon was equally adept for both broadcast and cable distribution.

In November 1980, Telidon's alpha-geometric coding became one of the three international standards recognized by the International Telegraph and Telephone Consultative Committee, the UN agency responsible for setting international telecommunications standards.[2]

The capability to generate and transmit superb graphics made Telidon teletext a very strong competitor from the time it was introduced. Many applications of teletext were identified by the Canadian Department of Communications. Some of them are listed in Table 3.1.

Above all, it was advertisers who felt that the Canadian protocol

**TELETEXT APPLICATIONS**

- News
- Advertising
- Theatre guides
- Astrology
- Classified ads
- Price comparisons
- Tele-education
- Emergency messages
- Bus routes
- Movie sub-titles
- Gardening hints
- Crossword puzzles
- Sports
- TV listings
- Restaurant guides
- Maps
- Best seller lists
- Grocery specials
- Computer games
- Computer art
- Emergency numbers
- Evacuation plans
- Cartoons
- Tourist tips
- Racing news
- Soil conditions
- Knitting patterns
- Sheet music
- Weather
- Program notes
- Travel schedules
- Real estate listings
- Tele-Gospel
- Government services
- First-aid information
- Tele-poetry
- Legislative news
- Community events
- Program captioning
- Job opportunities
- Quizzes
- Zoogeography

Table 3.1. Teletext Applications. Canadian Department of External Affairs—Ottawa.

was excellent from an aesthetic point of view. However, while Canada can be credited with substantially advancing the teletext protocol through incorporation of alpha-geometric graphics, no Canadian company ever prospered in the marketplace as a result of having such a competitive advantage. In Canada, most of the companies involved with teletext witnessed their greatest flow of money coming from the Canadian federal government rather than the private sector.[3] Canada felt very strongly about the future impact of Telidon on her economy and decided to do all in her power to make the technology a success. Unfortunately, rather than working hard to create and grow a market for teletext, many Canadian companies concentrated their attention on securing grants from the federal government.[4] This easy money lasted over such a long time that delays in making teletext a fixture in the marketplace soon made evident the fact that there would be no market. Canadians, like the Americans, saw their window of opportunity come and quickly disappear in a whirlwind of unresolved problems and issues. Their nature is expounded upon many times within the context of this book.

During the developmental stages of Telidon, two Canadian broadcasting networks offered teletext services. They were TV Ontario and the CBC television network. There was also another distance learning service at the University of Alaska and still another focused on government services at WETA-TV in Washington.

## TV Ontario

The first Canadian broadcaster (PBS) to initiate using the Telidon technology in the teletext mode was TV Ontario. They began offering their service in January 1980, mainly within metropolitan Toronto. Unlike other teletext services, however, TV Ontario concentrated on educational uses for their one hundred teletext pages. Within a couple of months they had a concentration of user terminals installed in six cities, eleven elementary and secondary schools, seven universities, four colleges, and three public libraries.[5]

By 1983, TV Ontario had expanded its service to include a career guidance system. This service offered ten-thousand pages of career and guidance information for students and guidance counselors. More specific services would follow; all made possible by a continuous flow of grants from the Ontario government.[6]

## CBC Broadcast Teletext

On 13 July, 1981, the CBC initiated its teletext service. CBC too used the Telidon technology and set up trials in Toronto and Montreal, and serviced a trial in Calgary via satellite. The overall cost of the CBC's project was about six million dollars, most of which was made possible through funding from the Department of Communications.

The CBC, like all service providers, was unsure from the beginning just what types of content would be most well-received. However, two areas the CBC felt most strongly about from the beginning were political reporting and dissemination of agricultural news.[7]

## University of Alaska Teletext Service

Although the University of Alaska wasn't a Canadian university, they did adopt the Canadian Telidon protocol for their distance learning broadcast teletext service. They started their service in 1983 and based it on the NABTS standard. Oftentimes we think of distance learning as an invention of the 1990s, but teletext had a major hand in the development of the concept many years before. The intention of the university was to transmit information or an entire computer program to their students. This was to be a convenience for those students who were unable to be present during the actual time that a course was being offered on television or by audio conference. They were able to access support material and discuss the presentation by calling up the course work instantaneously through the Learn Alaska Instructional Telecommunications Network.[8]

Given the vastness of Alaska and the difficulty of commuting to the university during the long winter, the university's teletext service was a forerunner of an educational service that was later to be recognized for its efficiency by all levels of education. Today, of course, distance learning has become a common technique employed in education.

## Alternate Media Center and WETA-TV Initiatives

Starting in 1981, New York University's Alternate Media Center pursued teletext testing in an attempt to answer to what extent it would be a useful medium for federal, state, and local government in making information available to the public. This approach would probably have made a lot of sense if the respective governments had been willing to provide grant monies to explore this question further. However, in the U.S., government never did see fit to become actively involved in facilitating the efforts of those who wished to determine a prominent role for teletext in the public sector. Money consistently was the name of the game, and without it there was no incentive to remain involved.

NYU was closely affiliated with WETA in Washington, D.C., a public television station, and was among the first to choose the Telidon version of teletext for their service.[9] During their trials they learned that hard news was perhaps the most popular feature on their service. And, in general, news, entertainment, and weather pages were among the most widely read. Armed with this information, in mid-1982, WETA shifted their emphasis in phase two of their trial to news analysis, community information, and video art.

To be more specific, they began to focus on the analysis of news and events in and around the capital. They used the Canadian Telidon standard and purchased most of their equipment from Norpak, a large Canadian supplier.[10]

## RADIO TELETEXT

It is perhaps worth mentioning at this point that broadcast teletext was not the only approach taken by researchers in Canada.

Some even managed to succeed in developing a radio form of teletext that was praised for its affordability to consumers and profitability for system operators. The system employed an unused
portion of the FM radio band to deliver text and graphics in color.
Receivers were priced at $150. Because FM sidebands were dereg-
ulated in the U.S. in April 1983, many investors were enticed to
invest in the technology. The downside was the fact that radio teletext was about five times slower than TV teletext, which itself was
very slow. Even at that, it was proposed that radio stations would
be able to generate additional advertising revenue with the new service for about fifteen thousand dollars.[11]

Somewhat prophetically, the inventors also talked about using
Direct Broadcast Satellite (DBS) for delivery of their service, which
they called Telefax. Today, of course, DBS is a dominant system
employed by information providers. Radio teletext never did become a prominent competitor to either the cable or broadcast version of teletext. What it did do, however, was add to the already
significant confusion plaguing those contemplating starting a teletext service.

## Major U.S. Studies, Surveys, and Speeches

Based on the assumed success of teletext in Europe, a number
of diverse interests throughout the United States began looking at
teletext with an eye to the future. Naturally, as with any new technology, there is always a significant amount of risk to contend with,
and any reassurances that could be gained prior to investing heavy
amounts of capital were actively sought through a series of studies.
Unlike in Canada, there was no direct involvement or funding for
teletext by the government in the United States. Consequently, it
was left to the private sector to start and develop the teletext industry. Because private enterprises must be risk-conscious in their
business decisions, there had to be evidence to support the merits
of financial investment in the fledgling teletext industry. This matter was addressed through studies, surveys, feelings of industry
leaders, and actual trials of teletext services.

## San Diego State University Study—1980

One of the earliest generic studies was conducted in mid-1980
by San Diego State University and entitled "Where Will Teletext Be

in 2000 AD?" The study followed rigorous research methods and techniques and employed the Delphi forecasting technique. Three rounds of questioning were conducted following strict guidelines and qualifications of participants. To determine the range of opinion, the researchers took the middle 50 percent of answers to all questions. Here is a summary of ten of the most relevant results they obtained:

1. It is very unlikely that the newspaper industry will decline as a result of competition from teletext.

2. It is somewhat likely that teletext will rely primarily on advertising to support its programming and service.

3. It is very unlikely that the availability of information on teletext will lead to a decreased emphasis on news and information by the commercial networks.

4. It is extremely likely that typical teletext receivers will have the capacity to store one or more pages in order to reduce waiting time.

5. It is likely that more than half of U.S. homes with television will have teletext receiving capability by the year 2000.

6. It is very unlikely that teletext services on cable systems will be mandated by the federal government.

7. It is very likely that federal, state, and local governments will use teletext to communicate public service messages and special announcements to the public.

8. It is likely that newspapers will be provided as a teletext service, using hard copies.

9. It is very likely that textual and graphic information to complement advertising will be provided via teletext.

10. It is likely that software for use in programming home computers will be distributed through use of teletext and videotex.[12]

It can be seen from these results that there was early awareness about many of the impediments that could damage or destroy the growth and success of teletext. Of course, with only limited historical information to guide them, it is somewhat speculative to assume that the participants in the study had a clear understanding or awareness of the role of teletext in the broader context of information dissemination.

## DONALDSON, LUFKIN & JENRETTE STUDY—1981

The teletext industry did not unfold in a business vacuum. Many well-respected consulting firms tracked the industry from the beginning and periodically published summaries of their observations and conclusions. Based on their international and domestic observations in late 1981, Donaldson, Lufkin & Jenrette offered the following conclusions:

1.  Teletext is likely to achieve widespread acceptance by mid-decade because:
    —It satisfies the consumer's appetite for information at low cost.
    —Transmission facilities (broadcast or one-way cable) are fully established already.
    —Teletext could be extremely profitable because costs of teletext-origination systems, operating costs, and programming costs are all very low.

2.  Cable teletext is likely to be more profitable than broadcast teletext because:
    —Cable channel capacity can accommodate several thousand pages of information compared with a few hundred on broadcast TV. This allows more extensive programming to attract viewers willing to pay for the service, and leaves plenty of space to incorporate advertising.
    —Consumers are already conditioned to pay for cable TV services. It is generally agreed that cable teletext could attract subscriber revenue, whereas broadcast teletext would have to be provided free.

3.  The industry structure most likely to emerge will involve several dominant national information providers who operate in partnership with local information providers and local conduits. The reasons are:
    —Initially, the financial resources of large companies are required for research and promotion. Furthermore, in order to drive down the cost of decoders, we need a rapid rollout using a standard set of equipment, a task that can be

achieved efficiently only by a large national system operator.

—Longer term, sophisticated local and national programming of uniformly high quality will be the key to success. Well-managed coordination of local and national information and advertising provided by many diverse sources will be necessary. It would be an overwhelming task for thousands of local information providers or systems to gather programming by themselves. Creating, collecting, editing, and distributing national programming and advertising to local conduits can be accomplished most efficiently by a one-stop national information service.

4. Time Inc. appears to be the company most likely to succeed because:
   —The company has positioned itself as a dominant national information and service provider that will work in partnership with local IPs and CATV operators. This is probably the correct model for the industry structure that will emerge in the future.
   —Cable-transmission facilities, which promise to be more successful in generating subscriber and advertiser revenues, have been chosen over the broadcast mode.
   —Financial resources for programming, R&D, and promotion are available.
   —HBO has given the company extensive experience as a national-program supplier working in conjunction with thousands of local CATV systems. The relationships the company has already formed will be valuable if and when Time Inc. attempts a national teletext rollout.
   —The product, which will combine information and entertainment, should be a good one. Sophisticated market research is now under way to determine as precisely as possible what the consumer wants. Quality control and attractive graphics are high priorities.
   —A national teletext business should produce enormous profits.[13]

Donaldson, Lufkin & Jenrette felt that if tests slated for mid-1982 were successful, a national rollout could begin in 1983–84, and a true national business could happen by 1984–85.

Looking for reassurance, it was felt that teletext could turn out to be a business that was run very much like HBO. However, the low programming costs and advertiser revenue suggested far greater profitability. Unlike HBO, a teletext business would pay no fees to program suppliers tied to subscriber levels, so operating costs were relatively fixed. Therefore, the revenue from each additional subscriber would be almost all incremental profit.

Sounding a cautionary note, however, the door was left open for the failure of Time's teletext project. Three major areas of potential difficulty were cited:

1. Coordination of information provided by thousands of local newspapers could prove to be a formidable task. Time may not be able to maintain quality control, which the company sees as critical to success.

2. The consumer may not be willing to pay for teletext.

3. Advertising, particularly national ads and display ads, may be ineffective in this medium. It is fast becoming the accepted wisdom that, in an interactive mode, commercial advertising must provide information. Classified advertising seems to be a natural intrusion, however, and image or awareness-building are not possible in a medium that has no sound and responds only to the specific requests of the viewer. Many advertisers are interested in the potential for "infomercials," especially retailers who have teleshopping in mind. The true effectiveness, however, is a major unknown at this time.[14]

## VIDEOTEX INDUSTRY ASSOCIATION SURVEY—1982

One of the most prestigious studies came from the Videotex Industry Association when, in early 1982, they published and distributed the "Teletext User Survey." This study was conducted by the Interactive Telecommunications Program at New York University beginning in October 1980 and focused on Ceefax and Oracle user reactions. While other similar studies were being performed at the same time, this one had a distinctly U.S. approach.[15]

The survey arrived at three conclusions:

1. Teletext has been well accepted by users despite the lack of an explicit need for such a service prior to its introduction. The primary keys to consumer acceptance appear to be the degree of convenience in receiving information in this way,

the ease of access, and the low cost of teletext receiving equipment in relation to normal television sets.

2. Considering the passive manner in which many subjects acquired their teletext set and the lack of a strong marketing effort for teletext at the time of the survey, teletext has reached a respectable number of users and will apparently continue to do so. Stronger marketing programs by the BBC, IBA, set manufacturers, and advertisers would likely promote further interest and users.

3. This indicates teletext has created a need for itself, at least within a portion of the British population. Teletext is then able to satisfy the needs and desires of those who use it, proving it to be an enjoyable and useful innovation.[16]

While elaborating on such important aspects of teletext in Britain as use of information, use of teletext, quality of service, feelings and opinions, and technologies associated with the two services, the general conclusion of the study was generally positive and encouraged initiation of teletext services in the U.S.


## DAVID M. SIMONS SPEECH—1982

By 1982 teletext had attracted the attention of every company and industry involved in the information dissemination business. Naturally their interest was piqued over the prospect of making a lot of money, but some were also concerned about the possibility of the new technology replacing their manner of delivering information. For these reasons there was a considerable amount of speculation and expert-opinion expressed to lend direction to teletext as a business.

One of the first guiding lights to express his vision of the future of teletext was David M. Simons, the highly influential president of Digital Video Corporation, a New York–based consultant to broadcasters, cable TV companies, and newspapers. In mid-1982 he expressed a number of opinions that had a marked influence on the composition of the teletext industry for the remainder of the decade. Simons felt that broadcasters would provide general-interest teletext services whereas cable TV would offer a service geared to special interests. This differentiation would occur, he claimed, because broadcasters are generally restricted to transmission of teletext on the vertical blanking interval, which can only accommodate a few hundred pages. On the other hand, cable can devote entire channels to teletext and provide access to five-thousand pages on each one.[17]

If broadcast and cable TV were to control the services, one might then wonder why teletext attracted the strong interest of the print media. Following the high attendance at the teletext-videotex session during the National Association of Broadcasters annual convention in April 1982, one of the most frequently asked questions was: Who is going to provide the information for teletext services? Since teletext was a written data service it was no wonder that what seemed exclusively a video service attracted the attention of the print media. Simons added to their interest when he made the point that staff members would have to be hired to provide copy to teletext services.

Simons was of the opinion that by 1989 at the latest, all television networks and their affiliates would be providing some form of teletext programming. He also predicted that most teletext transmissions would be pass-throughs of network service until a sufficient decoder penetration was reached. At that point, he felt, stations would begin to provide a local data service.

His prediction that teletext would develop the same way that color television was marketed—that is, the service would be provided by local stations and this in turn would stimulate the purchase of decoders by the public—never did materialize, and was always one of the difficult impediments to the growth of teletext.

Simons de-emphasized the importance of graphics on teletext because they took up too much of the limited page capacity. He also said viewers would not access a page which only featured a graphic. However, he expected to see greater use of high-quality graphics on higher capacity cable systems. He also felt that the fear that teletext viewers would zap newscast commercials was unfounded. "The best things teletext does best are fundamentally boring and extremely useful. There is nothing intrinsically entertaining about baseball scores or one's bank balance, but we will find time to get them and watch movies and ball games."[18]

Simons' comments were somewhat controversial because to some, graphics were teletext's key ingredient. Therefore, optimism still ran high that an inexpensive NABTS decoder would soon be available and give the competitive edge back to the broadcasters.[19]

By some accounts the cable version of teletext was more favorably positioned than that of the network broadcasters. Time Inc. especially seemed to have gained a lot of notoriety and will be the subject of chapter 5.

NATIONAL ASSOCIATION OF BROADCASTERS SURVEY—1984

A highly respected survey was conducted by the National Association of Broadcasters in 1984. By this time CBS and NBC as well as a number of cable TV networks had already begun to pursue teletext services of their own. At the time there was also a considerable amount of competition being felt from the print industry over who would succeed in providing the service most acceptable to consumers. However, since teletext was first and foremost part of the video signal it was only natural for television stations to feel it was rightfully their place to control teletext. For this reason it was important for them to see what their own professional association had to say about teletext.

Somewhat curiously, the study the NAB supported did not focus on NABTS teletext, but rather a sample of consumers who were provided with custom built Zenith television sets with a teletext decoder built in. Obviously, this meant evaluating a World Teletext Standard service. In some respects the study can be construed to have missed the point, because most of its membership was poised to pursue NABTS teletext. However, in spite of this puzzling approach, there was still something to be learned and shared about consumer preferences.

The NAB stated very early in their report that realization of a viable teletext service that would operate on a large scale in the U.S. depended on a number of interrelated factors. These included technological standards and capabilities, the degree of advertiser support, costs to the user, and perhaps most important, what services and information would be provided by the system. In their evaluation of teletext the study pointed out only one unfavorable feature of teletext; that being a decrease in favorable evaluation and viewing of teletext because of displeasure with the slow speed of access.

On the other hand, their study also identified several favorable aspects of teletext. The offerings named most often as "favorite" by the consumer sample were news and sports items. Fun page items and entertainment items were also frequently named. Overall, the favorite sections in the teletext system were in descending order: 1. News 2. Sports 3. The Fun Page 4. Entertainment 5. Weather 6. Consumer News and 7. Business News. With regard to the first two, a majority of news fans were older viewers (ages 40–49). Sports fans clustered in two age groups: 12–22 and 30–49.

The study resulted in three major findings:

1. The appeal of teletext appears to have had a long-lasting effect on viewing of the host station's local news program. Although the initial high level appeal of teletext decreased somewhat, early enthusiasts of teletext increased their viewing of the host station's local news program. The increased viewing persisted three months after installation, although teletext viewing and evaluation dropped by that time.

2. Teletext was used as a news medium and was related to television news consumption. The findings of this study, however, may indicate that the uses to which teletext is put reflect the strengths of the source of the teletext service.

3. There was evidence of a novelty effect for teletext for all demographic groups and family types. Regardless of personal or family characteristics, evaluation of teletext was lower three months after installation than immediately after installation.

The major implication for broadcasters from this study was the suggestion that teletext was perceived primarily as a news medium whose net impact was in a symbiotic relationship with the programming of the host station. Apparently, teletext increased one's appetite for news and in turn was reflected in higher viewership for the host's news programs.

Another major implication pointed out the likelihood that the threshold of news saturation had not been reached, as evidenced by the use of and inclusion of teletext into the heavy news-oriented consumer's routine. This void was advocated as an opportunity for increased success for teletext systems, since increased demand might be encouraged by keying offerings to augment, supplement, and complement broadcast news and lifestyle/entertainment offerings.

Finally, a third implication of the study was the probability that consumers' expectations for certain new electronic innovations would probably exceed the technology's capacity for performance. Impatience with "dwell time" (time it took for a full page to appear on the TV screen) and page turning, desire for more interactive capacity, and complaints about the quality of graphics were a few of the frustrations that accrued from teletext experiments and investigations.[20]

## STUDIES CONTINUE

Many other studies were done on teletext and for the most part it can be concluded that none of them matched or came up with the same set of findings as other studies. Their choice of variables to research was often different, data collection techniques varied, and results often matched favorable conclusions of previous studies. There just didn't seem to be a sufficient amount of focus on the downside of teletext. Consequently, service providers were left with the impression that the obstacles they faced were small and it was only a matter of time before they too would disappear.

Perhaps there were just too many studies to keep track of. As we have seen, studies originated from many diverse groups. As might have been expected, academia was very much involved with research. For example, in addition to San Diego State University (reported earlier), high profile schools such as the University of Southern California's Annenberg School of Communications also conducted studies.[21] There were many more, but in the end it appeared that service providers were more inclined to rely on their own findings and judgments.

## TELETEXT TRIALS PROLIFERATE

The interest expressed by American television networks and the print media in teletext was somewhat curious, inasmuch as the technology was devoid of U.S. roots. In early 1982, three protocols competed with each other to eventually become accepted as the U.S. standard. These included the British Context technology, Canada's Telidon, and the French Antiope system. Table 3.2 contains a listing of the major U.S. teletext trials conducted through 1982.

In the beginning it appeared that the British protocol would gain the initiative. After all, they were the first to appear on the American teletext scene. Since 1978, Bonneville, Utah's KSL television station had been running tests on the British technology. Furthermore, the British decoders were much less expensive and complicated than those of their competitors. Unlike the French and Canadian formats, the British system eliminated the need for a microprocessor to display information on the screen. These limited electronics at the decoder resulted in lower cost, better reliability, and a greater chance for success. However, the French became increasingly active in the U.S. and by 1982 had carved out a foothold. Among their inroads, they provided most of the equipment used by the *Courier-Journal* and *Louisville Times* in their teletext experiment in Louisville, Ky.[22]

Despite the KSL test, American teletext trials did not proliferate

## Major U.S. Teletext Trials

| Company | Project | Location | Media | Format | Start Date | Users/Sets |
|---|---|---|---|---|---|---|
| Bonneville | — | Salt Lake City | KSL VBI | Context | June 1978 | 12 sets, rotated |
| Field Electronic Publishing | Keyfax, Nite-Owl | Chicago | WFLD VBI and full-channel | Context | March 1981 | 40 in public places, 100 in homes |
| CBS | Extravision | Los Angeles | KNXT VBI | Antiope | April 1981 | 14 in public places and schools, 100 in homes |
| KCET | Now! | Los Angeles | KCET VBI | Antiope | April 1981 | 10 in public places, 40 in homes |
| New York University Alternate Media Center | — | Washington, DC | WETA VBI | Telidon | June 1981 | 14 in public places, 100 in homes |
| NBC | Tempo NBC Los Angeles | Los Angeles | KNBC VBI | Antiope | November 1981 | |
| Westinghouse | DirectVision | San Francisco | KPIX VBI | Antiope | 1982 | 30 rotated |
| Taft | — | Cincinnati | WKRC VBI | Context | 1982 | 40 in public places and homes, rotated |
| Dow Jones/ Danbury News-Times | — | Danbury, CT | Cable VBI | Antiope | 1982 | 100 in homes, rotated |
| Louisville Courier-Journal and Times | — | Louisville | Broadcast VBI | Antiope | 1982 | 35 in homes, rotated |
| Time Video Group | — | Orlando, FL/ San Diego, CA | Cable full-channel | Telidon | 1982 | 350 + homes |

(Chart compiled from a variety of sources by R. C. Morse)

Table 3.2. Major U.S. Teletext Trials. R. C. Morse—Marketing Communications Magazine.

until 1981. The presence of the three competing standards was readily visible in three major cities: British in Chicago; Canadian in Washington, D.C.; and French in Los Angeles.

## FIELD ENTERPRISES INITIATIVE

In Chicago it was Field Enterprises, Inc., publisher of the *Chicago Sun-Times* and owner of WFLD-TV channel 32, that became one of the early U.S. pioneers in teletext. They felt that the British Context system offered the most chance for success. Operating from this conviction, they participated in the start-up of a service they termed KEYFAX and began offering Zenith-equipped decoder sets to viewers. Examples of the configuration of some of their pages along with the service in general is described in detail in Figure 3.1.

Their late night program called NITE-OWL became quite popular. When it was first offered, Nite-Owl received a Nielsen rating of a 2.8 share, which translated to about seventy-five-thousand viewers who watched for an average of thirty to forty-five minutes a night. Nite-Owl achieved the early notoriety of becoming the only teletext service of its kind in the country, as well as the first to sell advertising space and time.[23]

## KNIGHT-RIDDER INITIATIVE

Knight-Ridder also became an early participant in the teletext competition. Like many competitors, they too concentrated their early efforts in a particular city. In their case it was Detroit, where in the early 1980s they expanded their cable press TV text after having successfully tested it in Lexington, Ky.[24]

## DECISION CONFRONTATION

In spite of this early documented success, the lack of consensus over the protocol standard in North America soon became a highly controversial issue. The British format featured an alphamosaic graphic capability, a fixed rather than a variable format, and serial rather than parallel attributes. This translated into the least expensive electronics combination to carry teletext to viewers. Function-

# KEYFAX™

KEYFAX is a new way to enjoy your television. Imagine choosing a subject from the exciting assortment offered by KEYFAX and having it displayed on your television.

You choose, and KEYFAX supplies, up-to-the-minute news, sports, business, weather, and leisure.

KEYFAX information is updated 24 hours a day, allowing you to access "key facts" at your convenience. Simply press a few buttons on the remote control handset. Within seconds, your selection is displayed on the television screen.

KEYFAX is America's first national teletext service. "Teletext" is the name given the system that puts words on your television screen. The teletext information is broadcast on an unused portion of the television signal and is displayed, in "page" format, on the television. One screen full of information is considered a page.

To receive KEYFAX pages you need a television set, a teletext decoder, and a handset. Within seconds, your selection is displayed on the television screen. Explore the various KEYFAX offerings by repeating this process for all desired pages.

KEYFAX is produced by KEYCOM Electronic Publishing. It was first introduced in Chicago in April 1981.

KEYCOM Electronic Publishing    Schaumburg Corporate Center    1501 Woodfield Rd.    Suite 110 West    Schaumburg, IL 60195    (312) 490-3200

Figure 3.1 KEYFAX Teletext Pages. Keycom Electronic Publishing—Schaumberg, Ill.

ally, however, this combination of characteristics did little to impress investors who were more interested in higher quality, especially in graphics. In retrospect, the French Antiope technology, which at first also featured alphamosaic graphics, offered only marginally greater flexibility at a higher cost, but where they excelled was in their ability to become the early favorite of the major television networks.[25]

## LOOKING FOR AND CREATING REASSURANCES

What compelled the major broadcasters to pursue teletext, in the words of David Percelay, early teletext project director of CBS, was a rare opportunity. "If we don't exploit what could very well represent the last great piece of broadcast real estate . . . then others with sufficient foresight can and will attempt to do so themselves."[26] What he was making reference to was the vertical blanking interval (VBI). However, in spite of this realization as early as 1978, CBS seemed committed to going with the state of the art they thought existed at that time. The French Antiope system was their choice, but later, as we will see, they had a major change of heart that dramatically altered their strategies.

One thing CBS did do correctly from the beginning was openly state that they felt the VBI was the property of local stations rather than the network. Gene Mater, senior vice president of policy, went on record to that effect when he told affiliates: "The company hopes teletext will serve as a broad-based information and advertising medium to complement the activities of its network stations and affiliates . . . as another means in which CBS may be able to deliver audiences to advertisers. . . . CBS hopes to market teletext pages as a 'real-time' alternative to advertisements currently being placed in local papers, thus not endangering network and affiliate broadcast revenues."[27]

What flowed from this statement was the clear identification of the two industries that would have to be properly dealt with: advertisers and the print media. Interestingly, the future of teletext was bound up, from the beginning, with nothing more than an extension of the competition between and among all information providers.

From the networks' point of view, teletext would be just like television—information-entertainment oriented. But what about the educational overtones to teletext? One of the major questions asked about television repeatedly has been the difficulty in using the medium more for education. Part of the problem has always been the reluctance of viewers to choose education when entertainment was available. However, this did not stop PBS stations from seizing on teletext as a special tool to enhance education. Early exposure of both teachers and students to teletext provoked a considerable amount of attention and a feeling that the interactive qualities of teletext provided an ingredient that was missing from regular educational television programs.

One major question that remained to be answered, particularly with regard to education, was whether people would be reluctant to read information from the teletext screen. The medium was, after all, silent; quite unlike television. According to Bob Geline, teletext project director for NBC, what they were looking for was "an answer to the basic question of whether or not people will read off the TV screen, and to gauge an initial measure of demand for the service."[28]

In an effort to make this determination, NBC had twenty-one national and local advertisers sponsor full and partial page ads to test the advertising potential of teletext. Graphics and color were heavy components of every ad. With this mindset, a commitment in kind was already being made to the alpha-geometric version of teletext.

## CABLE SYSTEMS GAIN SUPPORT

For a variety of reasons, cable television systems in partnership with local newspapers also got into the experiment and began to offer competition to the broadcasters. According to one cable system, 89 percent of their cable subscribers used the cabletext channel at least one time a week by the summer of 1981, an increase of 52 percent from the previous survey in 1980. Interestingly, most cable systems running teletext felt it was supplementing and complementing newspapers, not competing against them.[29]

Time Inc. was an early participant in cable teletext and seemed highly focused on how to position their service. They were also one of the most astute companies in assessing the status of the industry. According to Sean McCarthy, the Video Group director of development, "We're not launching a business so much as we're creating a consumer product."[30] However, when he later claimed that the average response time to a request for a page of information would be five to ten seconds, he stretched the truth. From the beginning the long wait time for requested pages proved to be a major impediment which teletext never did overcome.

One of the early observers of teletext, Marketing Communications, summarized the challenge very succinctly by making the point that those involved would have to be very much aware that a great deal of research remained before teletext became a legitimate medium for entertainment, information, advertising, and educational purposes. They also emphasized the fact that new styles of writing and editing, layout, and design were in the early stages of development and would have to be tried out, reformulated, and refined within each of the competing technologies' capabilities and limitations.[31] Apparently these cautions were not impediments or limitations, because 1982 was the year teletext was positioned to break out as a viable industry in the U.S.

## EARLY CBS AND NBC TESTS

The impetus for teletext to break out of its experimental mode in the United States may have occurred in July 1980, when CBS asked the Federal Communications Commission to adopt technical standards for broadcast teletext. Later, at Videotex '81, David Percelay, teletext project director for the CBS Broadcast Group, announced that following extensive engineering tests of various teletext systems, CBS had selected a modified version of the French Antiope technology for use in a program/audience test then underway at CBS-owned KNXT in Los Angeles. For the Los Angeles trial, Teledifusion de France loaned the project participants over one million dollars of Antiope teletext equipment.[32]

During the summer of 1982 both CBS and NBC succeeded in moving their interest in teletext from the experimental state to an actual plan of action. Both networks decided that the only way to truly determine whether a national service had a future was to start one.[33] Obviously they felt that it did because executives from CBS and NBC were quick to point out that affiliate interest was strong since it gave stations a real chance to go head to head with newspapers for local ad and classified dollars. In addition, two other revenue streams were identified. These were the opportunity for both national and local advertising revenue and an income stream from transportation businesses that might want to list their schedules on the service.[34]

Although NBC and CBS were firm in their conviction that teletext would become an asset for their programming mix, they still had to convince their network affiliates that the service was worthwhile. However, gaining their support was only one of many steps necessary before rolling out a national service. The major issues to be addressed included:

A. Selecting a teletext protocol

B. Acquiring headend hardware and software

C. Setting up a national page creation operation

D. Signing affiliates to join the network

E. Affiliates' willingness to purchase equipment for local page insertion

F. Signing up sponsors to fund the service

G. Acquainting viewers with the service

H. Getting viewers to purchase decoders to access the service

## CBS AND NBC's JOINT TEST

During the period from May through July 1982 the CBS KNXT-TV and NBC-owned KNBC-TV and KCET-TV (PBS) conducted a teletext research project involving seventy-five homes in Los Angeles. Meters were installed that measured channel selection and duration of viewing on the three stations, but measurement was confined only to homes in which the meters registered television viewing on any given day.

The two networks announced they would use studio equipment provided by Videographic Systems of America (VSA), a new American company dominated by CSF Thomson, the French telecommunications firm, and several other French companies. The cost to NBC alone was estimated to be between $250,000 and $1 million. This was a blow to the British World Teletext Standard and gave the French a competitive position which eventually, because of market forces, they had to share with the Canadians.[35]

Excluding a two-week start up period at the beginning of the test, when extremely heavy usage was recorded as people familiarized themselves with the system, the average household use of the services was 84 pages a day. Each page was viewed on the average for approximately 11 seconds for a total daily usage per household of 15 minutes. CBS-owned KNXT was most viewed, with an average of 36 pages per day per household, followed by KNBC with 32 pages and KCET with 16 pages.

There was some speculation that viewers would turn to their teletext service only at the beginning of a commercial break in news or entertainment programming. The results of the study proved this was not the case. An examination of this issue confirmed this conclusion: lead-in to teletext viewing, and television commercial activity during teletext viewing, posed no major concerns. Analysis of the data revealed that for those occasions where television viewing preceded teletext viewing, households left commercials to view teletext 20 percent of the time; 80 percent of the sessions came from regular programming. Thirty percent of the time viewers turned their television sets on specifically to first watch one of the three teletext services. Of this 30 percent, 60 percent went on to watch another program, while 40 percent switched off their sets after viewing teletext. Fully one-eighth of all sessions occurred when viewers wanted a specific piece of information, turned on their set to get teletext, and then turned it off again.[36]

Interestingly, on any given day the average teletext household only missed 4.5 percent of commercials they would normally have been exposed to had they not had the teletext option. With regard to viewing habits, access of all three services peaked during the 4:30 to 8:00 P.M. period. Usage fell during primetime and late night periods.

Viewers tended to look through large portions of specific subject categories. In order of usage, news, sports, travel, weather, business, and entertainment were the most viewed categories Monday through Friday. On weekends, sports dominated usage followed by news, entertainment, weather, and travel.

Local information tended to do very well in sports scores, weather, and travel and traffic information such as freeway congestion maps, ski conditions, and airline arrival schedules. Also, information offering rapidly changing comparison pricing, particularly of supermarket items and airline fares, was highly desired by viewers, as was information about sales and specials.

Viewers were particularly impressed with the continuous changing of editorial and advertising content and the ability to obtain desired information when they wanted it.

Most viewers in the test did not find teletext advertising to be intrusive or objectionable. In fact, some said they wanted more advertising, particularly informational advertising. The only objection was to ads that did not change frequently. Several viewers said they changed their buying habits because of advertisements featured on teletext, especially those having to do with specials at a particular neighborhood grocery store.[37]

In spite of the enthusiasm of NBC and CBS, ABC remained somewhat skeptical of the idea from the beginning. While they continued to explore their options they were not convinced that teletext was a cost-effective venture for them. And, unless the Federal Communications Commission gave its approval, the networks were stymied from making teletext available to viewers.[38]

One of the keys to optimism and continued pursuit of teletext was anticipation over formulation and acceptance of a North American teletext standard.[39] However, even without assurance of its ever becoming a reality, in the final analysis, teletext looked like too good an opportunity to pass up. It wasn't long before CBS, NBC, Time Inc. and a number of other well-known information dissemination companies converted this opportunity into their own business versions of teletext.

# 4

# Teletext is Born in the United States

IN SOME RESPECTS THE ENVIRONMENT AWAITING TELETEXT IN NORTH America was reminiscent of conditions surrounding the adoption of the NTSC color TV standard in the U.S., and of the subsequent competition between the NTSC and the PAL and SECAM standards worldwide. In France and Britain the issue of standardization of teletext systems was dealt with similarly. Both the broadcasting and telephone systems were government owned and controlled in France. The same was also true in Britain with the exception that there was an Independent TV Authority to govern broadcasting. While the content of TV programs was independent, the standards of transmission were strictly controlled.

Significantly, both countries adopted systems in which videotex and teletext formats were compatible, thereby making it possible to design inexpensive receivers capable of handling both teletext and videotex. In the United States and Canada, however, telephone, broadcasting, and cable TV operations were all predominantly in the private sector. This created problems because the teletext standards setting process took an abnormally long time to be resolved. What resulted was a situation where large service providers were inclined to commit themselves to a particular system somewhat prematurely.[1]

It was within this relatively unregulated open marketplace that teletext was greeted and nurtured. And it is not surprising that the British influence gained the earliest advantage in the United States.

## VIRDATA AND VIRTEXT

The first commercial, nonexperimental teletext service in the United States was the CableText system generated by Southern Satellite Systems. Their approach was both innovative and risky, but focused on being a leader. By later standards it was primitive, but it did enable information providers such as Reuters, Associated Press, Dow Jones, and United Press International to provide data signals over telephone lines to cable television stations. These signals were converted into synthetic video for incorporation into unused television channels. What the viewer saw was scrolling lines of text over which they had no control.

An important step in providing broadcast teletext was made possible when Ted Turner unveiled the concept of the "super station." Under this concept the signal of WTBS-TV channel 17 in Atlanta was distributed nationwide. This was done via satellite to earth stations, with the signal then displayed to local viewers. Somewhat later, WGN-TV channel 9 in Chicago also became a super station.

Based on their success in making the super station concept work, Southern Satellite Systems then conceived of the idea of using teletext to distribute data via the super stations's vertical blanking interval to its subscribers. This enabled cable stations to get around the heavy costs associated with leased telephone lines. Not surprisingly, the equipment used by SSS was based on the British teletext standard. However, it wasn't exactly the standard, but a precursor of that standard developed by Zenith called VIRTEXT. It wasn't until a page format of twenty rows of forty characters instead of twenty-four rows was adopted that teletext really replicated the British standard.

Zenith was probably the first American television set manufacturer to develop equipment for teletext. Following on the heels of the introduction of super stations, Zenith developed a small business computer system called the ZDS-89 which was programmed to create teletext pages and organize them into magazines. Later, Zenith went on to create the VIRDATA system which allowed for transparent data transmission simultaneously with transmission of teletext.

Unfortunately for Zenith, VIRTEXT and VIRDATA were specifically intended not to be compatible with a broadcast teletext standard. For this reason the major television networks turned their attention to other companies that catered to their specific needs.[2]

## KEYFAX NATIONAL TELETEXT MAGAZINE

Undaunted, Southern Satellite Systems carried on with their teletext plans and formed a joint venture between themselves and Keycom Electronic Publishing, which was itself a joint venture involving Centel Corp. (54 percent interest), Honeywell (30 percent), and Field Enterprises (16 percent). Centel at the time operated the nation's fourth largest independent telephone system, serving 1.2 million customers in ten states, and was involved in cable TV operations as well as the sale and distribution of business communications systems. Honeywell specialized in information processing, automation, and control. Field owned five UHF TV stations, the *Chicago Sun-Times,* the Field Newspaper Syndicate, cable TV systems, and interests in subscription TV and real estate development. They called the new venture KEYFAX. It became the first nationwide twenty-four-hour consumer teletext magazine on 15 November 1982, as a $19.90-a-month offering to cable subscribers. It was transmitted over the vertical blanking interval of the satellite service provided by WTBS in Atlanta.[3]

The immediate goal of the venture was to invest ten million dollars over the first several years and to attract two hundred thousand cable households within the first thirty months of service. To accelerate acceptance of their service, Keyfax offered cable operators $3.95 per subscriber signed. That amount was later to be increased to $4.95 per monthly subscriber when cable operators reached one percent of their subscriber base. Half the monthly subscription fee paid for the teletext service and the other half paid for the rental of decoders made by British suppliers.[4]

As pointed out earlier, SSS and Keycom had been experimenting with their teletext service in a free Nite-Owl Magazine broadcast on Chicago's WFLD, which aired daily from midnight to 6 A.M. And, to insure a steady supply of updated and relevant news, Keyfax contracted with news supply sources such as the Associated Press, United Press International, Dow Jones, the *Chicago Sun-Times* and Field News Service.[5]

The new service was managed by Peter Winter. Like his counterparts who headed up other teletext services, Winter was quick to comment on the progress of the industry. On one occasion he claimed: "America is a long way behind everyone else. At least 12 countries in Europe are broadcasting the British teletext system. It is an accepted way of life in Britain where between 600,000 and 700,000 people use it and 10,000 more are added each week. They've learned to use it as a kind of printed radio." He also stated: "What has slowed the growth of teletext in the U.S. is the myth that enhanced graphics capability is vital to getting information or an advertising message across a television screen."[6] With these two comments he touched on a couple of the most sensitive teletext issues in the U.S. Was the British or the NABTS standard going to prevail in the end? Interestingly, Winter felt Keyfax eventually would become totally ad-supported. But would advertisers be willing to spend money on what was primarily a text-based system?

## TELETEXT AUTHORIZED

On 31 March 1983, the Federal Communications Commission authorized the broadcasting of teletext services for reception by consumers. In making this announcement the FCC stated that teletext would not be subject to most of the rules and regulations of broadcasting such as the "fairness doctrine" and "equal time" requirements. While some broadcasters had hoped the FCC would choose among several leading teletext systems, the commission declined to choose, saying it favored an open market approach that would allow licensees the freedom of choice necessary to operate teletext services tailored to their own specific situations.

While the commission decided to permit both commercial and noncommercial stations to offer teletext, public broadcasters were concerned that the new technology would crowd out the closed-captioning signals then being employed for the benefit of deaf viewers.[7] What was particularly annoying to broadcasters, however, was the FCC's decision not to require cable operators to carry specific signals. This meant that cable systems could strip teletext signals from the VBI of the programs they were providing to their customers.[8]

## KEYCOM SHIFTS TO VIDEOTEX

By 1983, Field Enterprises was dissolved and its 16 percent interest purchased by Rupert Murdock's News America. Then, by early 1984, Centel had become the sole owner and a decision was made to downsize their teletext service. The association between Keycom and Southern Satellite Systems was discontinued and the teletext magazine originally carried on WTBS in Atlanta was picked up by Taft-owned WKRC in Cincinnati. The size of the magazine was reduced from one hundred to fifty pages, with the pages eliminated reverting to SSS, which used them for other business-oriented teletext information. At the same time Keycom switched their attention to videotex and by November of 1984 had launched a new videotex service. And, in June 1985, Keycom announced it was shifting its main attention to the development of a specialized business-oriented videotex service. Having made this decision, their break with teletext was now complete.[9]

## TAFT'S ELECTRA SERVICE

While Taft Broadcasting had picked up the remnants of Keyfax, they already had a teletext service of their own prior to this time. In June 1983, Taft and Zenith had just completed a one-year field trial of teletext on WKRC and were so satisfied with the results that they initiated the nation's first local commercial teletext marketing effort. Taft's president, Dudley S. Taft, proudly proclaimed, "It's historic because Cincinnati will become the first American city where the consumer can purchase teletext receiving equipment."[10] In this regard he was telling the truth since consumers could indeed purchase teletext decoders from retail outlets. As a point of fact, Zenith was the first and only American manufacturer up to that time to have produced such decoders. A picture of Zenith's system is depicted in Exhibit 4.1.

For a while it looked like the British World Standard had won a major battle, for at about the same time Taft and Zenith announced their venture, CBS suspended their teletext experiment and NBC cut their service back 50 percent. Their enthusiasm was heightened when Zenith announced they would be introducing an advanced and relatively inexpensive teletext decoder in 1984.[11] What wasn't emphasized too heavily was the fact that it would only work with Zenith television sets. However, to give Zenith the credit they deserve, they can be cited for a number of technical advances to teletext hardware which could have eventually become mass market items.[12]

The seemingly major disparity in fortunes could be traced to the difference in decoders needed to sustain the respective services. The Zenith teletext decoders sold for about three hundred dollars, while NABTS decoders, available only through cable and broadcast stations, were priced at a minimum of one thousand dollars. Even with this assumed victory, both Taft and Zenith had to be concerned over the fact that no other television set manufacturer in the U.S. felt strongly about getting into the teletext business. The competitors were, after all, participants in the same industry, and unless it could be made to grow there would be no industry in the future.

Taft reported that it cost them in the neighborhood of $185,000 to put their Electra teletext service on the air, and another

- Future digital product lines will include:
  - —More model selection.
  - —More total units produced in each model year.
  - —More features.
  - —More accessories.

All of these elements are part of the Zenith effort to promote World System Teletext. Join us on the leading edge of technology as we all work to make this medium a major success...

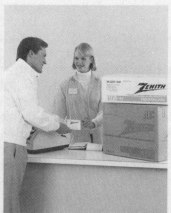

**Exhibit 4.1. A Zenith Teletext Receiver. Zenith—Glenview, Ill.**

$135,000 to cover annual operating expenses. While these costs did not vary markedly from what it cost to put a NABTS service on the air, Taft felt that future expenses could be contained by insisting on a tight writing style, offering only one story per page, and keeping stories to a maximum of eighty words. Furthermore, graphics on their World Standard Teletext system only required digitizing an existing photograph or artwork. It wasn't necessary to recreate graphics by hand on a computer light palette, as was required with most NABTS systems.

Electra's format was similar to that of a daily newspaper. There was an index for each section and the station had the capability of flashing updates over the lower portion of the television screen whenever a major news story developed. They even provided a clock and calendar which was always on display at the top of the screen. And, their decoders were also compatible with line 21 captioning. Taft seemed to have been convinced that the quality of the information they were providing far outweighed the importance of their graphics. However, this is not to say their graphics were that bad. They could be assembled quickly, simply, and inexpensively. But still a text mentality seemed to prevail. For example, on a full page, Electra could provide more than a dozen lines for an advertiser's message. An advertiser could divide the screen to accommodate up to thirty-six different products and prices. By comparison, an advertiser couldn't squeeze this much information into a radio or television commercial. And furthermore, the teletext ad was much cheaper.[13]

David Klein, at the time the TV critic for the *Cincinnati Post,* gave an unflattering appraisal of the service after having viewed it along with his family for two weeks. He claimed that at times it was more amusing than useful. He was quick to complain about the long time it took to access pages (up to fifteen seconds) and felt that local news was especially sparse and too preoccupied with police blotter items. He also thought Electra's artwork looked chunky and that their movie reviews were not very critical. Finally, he complained that feature stories sometimes didn't change for days. In response, Taft claimed that upgrading was planned for all of these areas to make them better.[14]

While they temporarily seemed to have gained some notoriety and sense of accomplishment, they did not manage to gain the competitive advantage they may have been looking for. In 1983, WKRC failed to gain and hold the upper hand in the teletext wars, but on the other hand, neither did CBS or NBC.

## CBS and NBC Prepare for Activation

Because teletext at the time was so successful in the United Kingdom, CBS and NBC were confident that it would also be well-received in the United States. However, before committing their financial resources to initiating a service, they had to first select a protocol. This was a major decision inasmuch as most of Europe was generating pages of information in which the graphics were alpha-mosaic with their characteristic jagged edges. This was aesthetically not very appealing. In the late 1970s the Canadian Department of Communications had succeeded in upgrading the protocol to the point where graphics were alpha-geometric. The ability to create arcs and curves helped make graphics more appealing. At the same time the French had achieved a similar breakthrough and were also offering alpha-geometrics through their Antiope service. At first both NBC and CBS were inclined to adopt the Antiope approach, however a breakthrough in the standards impasse eventually caused them to move in a slightly different direction. Part of this adjustment can be explained by exploring some of NBC's strategy.

## NBC's Strategy

In April 1983, NBC became the first broadcast network to produce and transmit for demonstration purposes a full high-resolution teletext service, using the North American Broadcast Teletext Specification (NABTS).[15] This was a major move. According to Barbara Watson, general manager of NBC Teletext, "High resolution graphics capability is one of the key factors in creating an advertiser-supported teletext service. . . . This will allow NBC to present more accurately company logos, product displays and creative graphic elements, which NBC considers essential in the development of an advertiser-supported service."[16]

The following month NBC also became the first broadcast network to put on the air a NABTS national teletext service. In making the announcement of initiating their service, Watson stated: "We did not want to be premature. We felt that high-resolution graphics . . . would be essential to make teletext a viable business opportunity. We believed so strongly in the capabilities of high-resolution teletext that we were unwilling to put older technology (called alphamosaics) on the network. We chose to wait until the new technology was ready."[17]

Then, in June 1983, speaking at Videotex '83, Watson proclaimed, "Consumers are looking for information that touches their lives, whether it helps them live better or makes them feel that they're in the know. . . . The key words are concise, timely, convenient and selective. . . . Our challenge is focus. How do we bring to the largest audience base exactly what they want? Is teletext a technology in search of a market? Yes, it has been, but also, yes, it will find its market."[18]

With this in mind, NBC Teletext contained eighty pages. After a magazine-like index page, which also highlighted the array of daily teletext options, viewers could choose, in the words of NBC, from the following features:

NEWSFRONT—This section is even more timely than today's newspapers in its presentation of the key international and national news stories of the day, as they happen. Stories of importance are being updated, as fast as they occur.

WEATHER—A national weather map is providing an instantaneous update at the touch of a button.

SPORTS—Not only is NBC Teletext reporting today's scores, but it is also furnishing sporting news and feature stories that are designed to capture the imagination of both ardent and casual sports fans.

MONEY—NBC Teletext is bringing the viewer pertinent business news and timely features on personal finance, in addition to up-to-date news on the fluctuations of the stock markets.

PEOPLE—NBC Teletext's "people" page is reporting on the comings and goings of interesting people in today's news.

YOUR BODY—NBC's electronic magazine is helping to keep its viewers in shape by offering health tips and animated exercise sequences.

LIVING—This section, aimed at the daytime viewer, offers tips to the homemaker.

HOROSCOPES—This feature is designed for those who rely on the stars for their answers.

SOAPS—NBC Teletext is offering synopses of last week's daytime-drama episodes for those who have missed them, and is providing "teasers" on current NBC-TV daytime starts.

KID'S KORNER—Aimed at younger viewers, "Kid's Korner" answers youngsters' questions with material which is both informative and entertaining.

PARTNERS—NBC Teletext has designed the industry's first high-resolution teletext romance serial, which employs a "cliff-hanging" style geared to bring viewers back the next day.

TRAVEL—NBC Teletext is offering an informative travel catalogue, as the first in a series of catalogues.

MOVIE REVIEWS—Picks and pans of current offerings on the silver screen.[19]

In Exhibit 4.2 is an example of a typical index page used by most teletext service providers. You will notice the number of the page the viewer had to press on a keypad in order to access a particular subject.

Perhaps this assemblage of information was precisely a reflection of what Watson had repeatedly said about NBC Teletext. However, it wasn't too many weeks later before both the NBC and CBS services were put on the verge of termination, mainly because inexpensive NABTS decoders were unavailable and consequently the

Exhibit 4.2. Index Page Sample. WIVB-TV—Buffalo.

viewership for teletext on the networks was virtually nil. The momentum, for the time being, began to swing back in favor of the World Teletext standard. However, after weathering the aforementioned scale-back in their services, as described in chapter 3, both networks began working on plans to come back strongly in early 1984.

## 1984—The Year of Promise and Hope

In mid-1983, CBS suspended its fledgling teletext operation EXTRAVISION for two months and NBC temporarily cut back its teletext service to fifty pages from one hundred pages. Both withdrawals were blamed on the slowness of television set manufacturers to build inexpensive decoders. After having spent a combined fifteen to twenty million dollars on their respective teletext service development during the prior four years, both CBS and NBC were compelled to become more budget oriented.

At the same time, the FCC's refusal to require cable operators to transmit the networks' teletext services as part of their service to subscribers also had a negative impact on the plans of both CBS and NBC.[20]

However, encouraged by their own tests and those of others, the teletext services of CBS and NBC were ready to launch in 1984.[21] A great deal of enthusiasm was generated by an announcement very early in January 1984 by Panasonic that they had started manufacturing a NABTS terminal (see Exhibit 4.3) and planned to begin marketing it in Charlotte, N.C. by the end of the month.[22] Although this terminal was intended to be only the first of two interim devices until a Very Large Scale Integration single-chip decoder could be introduced in 1985, the last barrier to the networks introducing a teletext service was now removed. Unfortunately, there was still no mention of the price for the terminal.

Sony and Sharp were similarly reported to be developing VLSI decoders for NABTS teletext. Promisingly, these were to be built into new television sets and incorporated into separate decoders by April 1985. In the meantime, Panasonic planned to sell their terminals at a loss as a market experiment.

The inability to determine a price for the terminals created many complex problems for all concerned. You can't blame companies for wanting guarantees to amortize their R&D efforts, but what the industry really needed badly at this stage of development was a clear commitment on the part of hardware vendors.

**Exhibit 4.3. Panasonic Teletext Decoder. Panasonic—New York.**

This meant that acceptance of teletext would be slow. Furthermore, there was still no standard system yet and great uncertainty over whether local broadcasters, advertisers, and the public would want teletext. According to Gary H. Arlen, publisher of International Videotex-Teletext News, "People are still not sitting at home hoping they'll have teletext."[23]

Closely on the heels of the announcement by the major hardware vendors that a new decoder was available, both CBS and NBC were now positioned to finally make their move.[24] After each had invested approximately fifteen million dollars designing and testing the technology, they committed another three million each in 1984 to offer one hundred pages of text, updated every fifteen minutes.[25]

## CBS AND NBC ROLL OUT TELETEXT

On 4 April 1984, four days after the decision by the FCC authorizing teletext transmissions, CBS began the first national over-the-air electronic information service. Even though decoders were very expensive and in short supply, Albert Crane, CBS vice president in charge of Extravision, remarked: "It's just like color television in the 1960s. The networks began broadcasting in color to help develop that market even though there weren't many color receivers. We're driving a new service and getting people interested."[26] Even though most affiliates were now carrying the teletext signal in the VBI, the service was virtually invisible because of the unavailability of inexpensive decoders.[27]

To some observers, this was a modest start. However, Crane and Barbara Watson, in charge of NBC Teletext, stated that this was exactly the speed at which they were comfortable to move ahead. Perhaps they were mindful of the demise of Time's cabletext operation, which was reported to have spent between seven and nine million dollars on their service during the previous two years before dropping out of the industry.[28] With only a little over 40 percent of the country wired for cable and confronted with having to negotiate with individual cable operators hard-pressed for spare channels, Time was up against too many major obstacles to make their initiative work.

The bad experience by Time apparently had a negative influence on ABC, for they concluded from the very beginning that they saw no future in the service. In addition, up until Panasonic's announcement, decoders were still priced in the nine-hundred-dollar range and industry analysts predicted that it would take three to five years before this price could be lowered to the one hundred dollar level, the figure believed to be needed to give teletext a mass appeal.

In spite of the difficulties, CBS was highly optimistic since they claimed they were being supported by 167 (85 percent) of their affiliates; a figure that represented seventy-one million homes. NBC was similarly optimistic.[29]

## WBTV-CHARLOTTE

On 4 April 1984, CBS made the announcement that "Extravision" had been launched on WBTV, the Jefferson-Pilot station in Charlotte.[30] The service provided one hundred pages of information, twenty generated locally and the other by the network. National pages were generated twenty-four hours a day by a staff of twelve based in Los Angeles. The first page of the service provided the most important headline of the moment and gave a page number where more details could be found. Beneath the headline was spread a full index of the day's total feed which the viewer could select from. Most pages of information included news, sports reports, health, science, and communications features, food tips, and advertisements.[31] For creation of their local pages and insertion into the VBI, WBTV used a system manufactured by Videographic Systems of America (see Exhibit 4.4).

The station stated that it would cost them approximately two hundred thousand dollars a year to run, a figure arrived at even before they had talked to advertisers or decided how many local pages it would offer and how much it would charge. With prospects of starting their service in April 1984, the station hoped to have between two and three thousand decoders available for sale. Their enthusiasm was heightened when Panasonic announced that they would make these decoders available at nine outlets in Charlotte at a three-hundred-dollar promotional list price rather than charging the going rate of nine hundred dollars. Panasonic also paid WBTV forty thousand dollars for advertising, a tack that strengthened the belief that teletext would easily be supported and become profitable through advertisements.[32]

Although it was not widely pointed out, the decoders could function only with TV sets that were equipped with an RGB (Red-Green-Blue) connection. Only a small number of standard sets at that time had this connection available. To address this problem,

**Exhibit 4.4. VSA Page Creation System. Videographics Systems of America—New York.**

Panasonic at the time was also selling a twelve-hundred-dollar list price teletext monitor in Charlotte.[33]

Although both the CBS and NBC services were advertiser-supported, free to the consumer, the CBS Extravision launching in Charlotte was seen mostly by customers in television stores, since almost no decoders were yet in the hands of consumers. Undaunted, one of the station's executives reminded those assembled for the launch of the service that in the early days of television, sets were sold at public demonstration sites.[34]

Closely behind in announcing plans to offer a teletext service were affiliates in Buffalo and Seattle (the Buffalo experience is the subject of chapter 7).[35] Interestingly, the bigger stations such as WCBS-TV in New York City and KNXT-TV in Los Angeles had no plans to come aboard until 1985 or later.

On the very day WBTV began offering a teletext service, CBS, NBC, and Group W combined forces to announce they had finally committed to the NABTS protocol. The issue of competing standards was now reduced to two choices.

On the opposite side was Taft Broadcasting, which was operating a teletext service in Cincinnati based on the British World System rather than the NABTS.[36] Indeed, in 1984 there were still three competing teletext protocols that had originated respectively in Britain, Canada, and France. This added to the indecisiveness of every business associated with teletext. While the FCC authorized teletext transmissions on 1 April 1984, the agency also ruled at that time that it would leave it up to broadcasters to decide what system they wanted to use. Needless to say, this lack of consensus in standards had a divisive effect on the industry.

## CBS AND NBC POSITION THEIR SERVICES

While CBS took a market-by-market approach, NBC positioned itself to feature major events where a show-and-tell kind of operation rather than a commercial enterprise was to be featured.

Of course, the key to the financial success of the service hinged on advertiser support. However, according to industry experts, that would not happen until national penetration reached the two to three percent mark.[37]

At the time of launching Extravision, CBS also announced that it was expanding its existing captioning efforts for the hearing impaired through introduction, on a transitional basis, of dual mode captioning. This was a system in which simultaneous captioning was provided in two different formats—the so-called "Line 21" system and the NABTS. This enabled hearing-impaired viewers to receive closed captions through decoders compatible with either format. As a gesture of goodwill, CBS donated several dozen decoders for Charlotte hearing-impaired community use.[38]

Optimistically, Albert Crane announced that CBS saw a 10 to 12 percent penetration by the end of 1985 or early 1986, amounting to nine million households. Even more optimistically he foresaw decoders priced under one hundred dollars built into all new television sets within three to five years. Furthermore he claimed that local origination was where the real future of television teletext would lie because of viewer demand for local news, weather, and sports.

By early May, CBS affiliates were being encouraged to support Extravision with local pages of their own. Since most were already passing the signal, the next step was to offer local demonstrations. This cost was estimated to be between $30,000 and $40,000. In order to offer local pages it would have been necessary for each station to invest approximately $200,000 for equipment and another

$100,00 to $125,000 each year for operating expenses. Obviously, the only way this type of investment could be justified was to attract advertisers willing to foot the bill. However, advertisers function on the basis of market penetration, and without a large viewer base it was difficult for stations to entice very many of them to invest in the fledgling service.[39]

Nevertheless the hype continued. The list of companies reported to be developing the new inexpensive decoder grew to include: Quasar, Hitachi, RCA, GE, Honeywell, Magnavox, Zenith, and Samsung, in addition to Panasonic, Sony and Sharp.[40] In May the standards issue also heated up. CBS was quick to show the superiority of NABTS graphics over those of the so-called world standard. The underlying strength of this protocol appeared to be the ad agencies' insistence on optimum quality. With dollars so scarce the competition naturally became more intense.

On a positive note, CBS announced that KCBS-TV in Los Angeles would be demonstrating Extravision at fifty locations beginning in early July. In addition, an announcement was made that the Bonneville station group, one of the pioneers in teletext experimentation, had agreed to participate in the Extravision service with Bonneville's CBS affiliates.[41] This included the Pacific Northwest and Mountain West areas. Bonneville's decision to convert to the NABTS standard was a major shift for them. They originally had been using a version of the World System Teletext since 1978 that was strangely incompatible with the one being used by Taft's WKRC in Cincinnati.[42] When operational, the stations called their Teletext-5 service a "video paper," which in most respects resembled the local services of both WBTV and WIVB. Bonneville claimed that a TV station could put a teletext service on the air for under one hundred thousand dollars. However, one is led to wonder if the same equipment was being used for both KIRO in Seattle and KSL in Salt Lake City. If so, much of the costs would have been shared, consequently diminishing the amount spent per station.[43]

By November 1984 only three CBS affiliates were on the air with local teletext magazines: WBTV in Charlotte, WIVB in Buffalo, and KCBS in Los Angeles. Only one NBC affiliate, KNBC in Los Angeles, was producing local pages by that time. ABC, which shunned teletext, instead was spending its money on three other programming areas. These included an 80 percent interest in ESPN and one-third interest in both the Arts and Entertainment network and the Lifetime network. According to their plans, they could always start a teletext service at a later date.[44]

To further entice affiliates to make the investment so they could generate local pages, CBS was quick to point out that there was a need for local classified ads, graphic portrayals of local traffic conditions, airline flight arrivals, and up-to-the-minute information about entertainment available in the market. In addition, it was suggested that a station could supplement local newscasts with background reports and sidebars, and local advertisements with details on sales and specials. The obvious links to advertising revenues were clearly emphasized. Pressing the issue further, Crane claimed that Extravision had been designed as an affiliate service from the start. "It can establish your station as the innovative technological leader and improve your overall community services as well as help maintain the vitality of over-the-air broadcasting by enabling stations to compete against textual services provided by hard-wire distributors."[45]

## PROJECTED USES OF TELETEXT

Many novel uses for teletext were envisioned. According to Crane, "I envision the day when major sports events might have a chalkboard behind the screen within the blanking interval for more information on the play the viewer has just witnessed." He also predicted that in the future, as a breaking news story unfolded on the "Evening News," the Extravision page number would appear on the screen to indicate that printed details were available at the press of a button.[46] Coincidentally, this is exactly what many television stations are suggesting now, but instead of teletext the printed details appear on the World Wide Web.

In spite of its superior graphics, NABTS still was unable to gain a competitive advantage over other teletext "magazines" including Keyfax, the World System service in the lead in the U.S. in terms of subscribers. Furthermore, the World System, based on the British Ceefax technology, had the major advantage of using home terminals that retailed for just three hundred dollars.

Not surprisingly, the group most interested in the demonstrations to date were mainly deaf persons, clearly more favorably impressed with the closed-captioning feature of the services.[47]

There can be no denying that teletext was viewed as a potential big moneymaker. The entire configuration of teletext was geared with that thought in mind. Group W was among the first companies to focus exclusive attention on how teletext could deliver more service and profits. They developed what was termed DirectVision, the first large-scale test of electronic classified ads in North America.

The concept was refined in a field trial they conducted in San Francisco. They provided the viewer with three separate magazines that appealed to different viewers' needs and interests. One of these was termed The Shopper, a quick and easy guide to the day's special buys and bargains being offered by many stores and retail outlets in the Bay area. Still another magazine was named the Metro Mart, a summary of all the classified listings from several newspapers. And finally, their third magazine was called Newsline, a fast-breaking news, sports, business, and weather synopsis. To call attention to these services, Group W invented the "DV EXTRA," which consisted of a small "DV" superimposed over some television programs and commercials. These called attention to the fact that additional information relating to the program or commercial was available if the viewer switched to DirectVision. Very often a viewer who accessed the "DV EXTRA" logo was rewarded with a listing of the day's best bargains and prices at specific stores.[48]

Group W's approach seemed to fascinate the media. They covered their initiative in considerable detail, but even they recognized the fact that the standards competition and lack of integrated TV decoders were becoming major difficulties for the growth of teletext.

## FOCUSING ON BUSINESS

Because revenue generation was so difficult, it was only natural that there was at times a surge of interest in using broadcast teletext for the dissemination of business data. There was competition for delivery of such services from FM sideband and DBS satellite. However, what all wireless systems shared was the fact that both senders and receivers of information could save considerable sums of money by avoiding the use of long distance or leased line telecommunications charges; the primary means, at the time, for business data to be forwarded.[49]

Among the business customers most sought after were the banks. They did, after all, have plenty of money and were interested in improving the speed and efficiency of banking transactions. However, their preference was two-way videotex services. Nevertheless, some industry analysts saw a role for teletext in home banking. While it was not envisioned to be a stand-alone system, it was felt that a type of hybrid system with videotex was a distinct possibility. For the most part, teletext pages were to be used to send a complicated series of instructions that ultimately prompted the customer to go to the videotex mode where actual transactions could be recorded.[50] For many reasons, this configuration never gained interest or acceptance.

Still another hypothetical business use of teletext was explored by Merrill Lynch and the Public Broadcasting Service (PBS) to use teletext to provide investment data to Merrill Lynch clients. To test the interest level and efficiency of this approach to investment information, data was sent to terminals in locations like public libraries rather than to TV sets in private homes. This maximized the use of the limited number of terminals then available for tests.[51] However, this approach to accessing investment data also failed to impress consumers.

With the introduction of teletext it became clear that television was going to become digital rather than analog sometime soon. For their part broadcasters were not quick to make this change. Individuals and businesses who accessed data, on the other hand, were ambivalent regarding the means by which the needed information reached them. Because satellite did not have the terrestrial problems often associated with local television transmission, it already had a superior quality that strengthened its case for being the favored technology to deliver business information and data. A number of companies pursued this approach and many survive to this day, which is more than can be said for the services that attempted to utilize broadcast teletext.[52] For a more in-depth review of the NABTS protocol and the marketing strategies of CBS and NBC, see Appendix 2.

Because of its being a content-driven information service, teletext not only appealed to broadcasters and cable system operators, but also many of the largest print companies in the U.S. Time especially was a major NABTS proponent and according to at least one study stood the best chance of any company to succeed with their teletext operations (Donaldson, Lufkin, & Jenrette study). Their strategies and experiences are the subject of chapter 5.

# 5

# Print Media Becomes a Major Participant

TELEVISION STATIONS WERE NOT THE ONLY PARTIES INTERESTED IN teletext. Several print media also had aspirations of creating and profiting from a teletext service. For example, the Courier-Journal and Louisville Times Company conducted a cable teletext test of 150 households in Louisville and Jefferson County in early 1982. Somewhat later that same year Knight-Ridder Newspapers announced that it would expand its "cable press" TV text into the Detroit area from Lexington, Ky. Their affiliate, the *Detroit Free Press*, entered into a joint venture with two suburban cable companies to provide a twenty-four-hour service built around news, sports, and classified ads.[1]

The anxiety being experienced by the print media was periodically reinforced by reports issued by several notable research firms. One such example was a report issued by VideoPrint, a newsletter published by International Resource Development, when they stated that teletext news services threatened to drive evening newspapers to the very edge of survival because teletext could be updated instantly.[2]

## TIME ENTERS THE FIELD

The most notable print company that became involved with teletext was *Time* magazine. In the early 1980s, Time was the nation's largest publisher of magazines and books. Among the more popular magazines they published were: *Time, Life, People, Sports Illustrated, Fortune, Money*, and *Discover*. In addition, because they owned numerous cable television stations they envisioned a national teletext service operating through these outlets.

Time approached teletext from a position of strength. In the early 1980s they were the country's largest publisher of magazines and books. Their strategy was to become an even more prominent major supplier of information for the consumer marketplace, by distributing their service through local cable operators. They were prepared to focus on content. Their expressed interest in the technology was relevant only because it affected the distribution and presentation of the content they were offering.

In early 1981, Time decided to use TELIDON for the first nationwide, multichannel teletext service for home cable subscribers in the U.S. They had 1.5 million cable subscribers on their American Television and Communications Corporation subsidiary, and through another subsidiary, HBO, had access to another 4.5 million cable subscribers. In making the announcement of their selection, Sean McCarthy, director of the Time Inc. Video Group Development unit, said: "After reviewing all the competing teletext technologies, we determined that Telidon is the most desirable because it allows the greatest degree of editorial flexibility. Its capacity to produce graphics exceeds the current capabilities of other teletext formats."[3]

## TIME ASSESSES THE MARKETPLACE

The control center for Time's teletext service was termed the Consumer Communications Center, located in the Time and Life building in New York City. It was a highly sophisticated research facility that was created specifically to facilitate the development of Time's teletext service. Like most new business ventures, efforts are made to minimize financial risk by first conducting extensive research on the feasibility of the innovation. The approach taken by Time mainly featured reliance on information gathered from focus groups.

Approximately one thousand persons of all age groups viewed Time's teletext service and participated in focus groups at the center. All were paid volunteers, contracted through the use of a professional screening service. During the testing process, users were first informed about the teletext medium and Time's teletext service. They then spent time alone in living room–like settings where they had hands-on experience with the teletext service. In the focus groups which followed, they discussed their responses to the content areas while Time staffers viewed them through a two-way mirror in the lab.

Four major conclusions were arrived at through focus group testing:

1. Teletext is not a medium for intensive reading. It is a dynamic, dip-in service which users will go to for specific information needs. It has to be up-to-date, spontaneous, useful information, as well as entertaining.

2. Information must be easily accessible. The menu and tree-structure organization must be "user-friendly."

3. Interactivity is highly successful for holding users' interest. Games and quizzes are especially appealing.

4. Graphics should be used mainly to convey information or add a pleasing note to a screen.[4]

## TIME'S STRATEGY UNFOLDS

In February 1981, Time, Inc. announced that it would launch a nationally distributed teletext service by year's end. This was to be done through Time Video Information Services, Inc., a newly formed subsidiary of Time, Inc.[5] A full video channel teletext service was developed, but it was not until the fall of 1982 that trials were conducted in San Diego and Orlando.[6]

Originally it was anticipated that Time Teletext would consist of up to five thousand full-color pages of information and entertainment available to viewers around the clock. Both national and local news were to be featured, and Time reasoned that because of its expertise in national information delivery they were favorably positioned to bring customers a high quality service. And, consistent with their print media mentality, Time decided that newspapers, which they considered the best sources for local news and information, would provide the local input.

Revenue to sustain the service was projected to come from advertising and a fee of between five and ten dollars per month from cable subscribers. To cover the nation quickly with their teletext service, Time planned to transmit their signal via satellite to cable operators who in turn would distribute the service on one full video channel to subscribers.

In 1982, some two years after the San Diego State report, Time provided some valuable insights of their own. According to Peter Gross, general counsel to Time's Video Group and chairman of its recently formed Video Information Service Corp., "We're all involved in a long learning process. Nobody really knows what people will want. This company has managed some of the things we have out of a combination of good luck, planning and just being in the right place at the right time. We can't get seduced by the technology. The consumer doesn't. We're all involved in a long learning process."[7] While his view appeared to be cautionary, Sean McCarthy, director of development for the Video Group and president of the new corporation, emphasized the business side when he stated that Time expected its new venture to be a "business of scale"— that is, returning one hundred to two hundred million dollars in revenues—"for the last half of the eighties."[8]

The key to success was generally agreed to be the methodology by which information was to be disseminated. Don Sider, managing editor of Video Information Services and perhaps the top journalist among the planners at Time, stated, "Its like a sculptor chipping away all the marble that isn't a human fact. It's slow going." He went on to say that "we get a much heightened sense of the someone on the other side of the TV screen. They're not creative. You can't ask them what they'd like to see. But they sure are critics. They know what they don't like to see."[9] However, by early 1982, Time reached two important conclusions: one, "retrieval that takes longer than 10 seconds is unacceptable. And two, graphics are fascinating the first time, boring the second, and actively irritating the third."[10]

Since most of Time's services group was composed of journalists, they worried a lot about becoming preoccupied with that they termed "print-think." According to Sider, "Every day we snip the old ties and make up a teletext way to do things."[11] Among these ways was a branching mechanism that compelled or permitted a user to choose among three things: interaction through quizzes by which a correct answer was a gateway to a new page; graphics that moved; and immediate access to and updating of information otherwise unavailable until it was printed and distributed.

In spite of all the obstacles and need to invent, strategies did take shape and provide guidance for the future. One of the most important of these was the plan to ultimately have local news inserted into the service after reworking by teletext editors. With this in mind it was assumed that future alliances would be exclusive in communities with more than one publisher. In addition, the papers would share some undisclosed amount of the subscriber revenues, unlike other suppliers of hardware and software who got flat license fees.

Far and away the most disturbing impediment to strategic planning was the dismal state of the hardware technology. For example, in 1982 system headends were scarce and available only in prototype. They were also very expensive, costing just under one hundred thousand dollars. It was feared that this would scare off operators and impede the further development and expansion of teletext. At the other end, set top converters posed their own set of problems. While it was assumed that subscribers would pay nothing for access initially and then five dollars a month later on, no mention was made about who would pay for the box, whatever its eventual cost.

The view of the future from 1982 firmly established Time as a major participant in the information services business. "Quite apart from teletext, we're looking into full participation in electronic publishing," said Gross.[12] These amounted to two major dimensions: upgrading the teletext plan and serving as an information supplier to others.

"We're in a constant process of projecting how common computer-driven information and entertainment centers will become in the home," said Gross. "Will there be a reliance on storage, interactivity and personal computer packages? If so, we want to fit in and feed other elements of that mix with a broader information base, with the offer of transactional services, with software to be manipulated by a home computer."[13]

## TIME RETHINKS ITS STRATEGY

In March 1982, Sider surprisingly made the statement that "teletext will never replace print publishing."[14] This succinctly revealed a company at loggerheads with itself. If teletext didn't at least threaten to cut into the print media, what was Time doing spending so much time and money on this new electronic technology?[15]

At the National Cable Television Association's conference in Las Vegas in May 1982, Larry Pfister, vice president of Time's Video Information Services, elaborated more fully on some of the questions Time was still trying to resolve. While it may have been tempting to conclude that teletext was already a done deal, Pfister's comments expressed a more cautionary note:

> . . . at TIME Inc. we are careful to temper our basic optimism with the reality that not every technological breakthrough is automatically predestined to become a marketplace success. . . . We're trying to find the

real or perceived value "Joe Six-pack" actually sees in using his television set for information services.[16]

Time did feel at the time that their teletext service, which was capable of transmitting up to five thousand pages of information continously using the bandwidth of a full cable channel, offered many competitive advantages. When compared with using the broadcast vertical blanking interval, a full-channel approach had about fifty times more capability to offer. This being the case, the only question remaining was whether audiences had a need for this many pages of information.

Time felt that the cable industry was better positioned for text services, for technical reasons, than either the broadcast or telephone industries. According to Pfister:

> Technically, cable already offers a broadband communications delivery system into nearly one-third of the households today, and penetration is increasing at a geometric rate. . . . Broadcasters have only the Vertical Blanking Interval of a television signal, or sub-carrier of FM radio, to transmit text, severely limiting the amount of information they can carry.
>
> Further, it is doubtful whether broadcasters will jeopardize their primary business by permitting teletext competition for their commercial viewing audience. . . . The telephone companies may have broadband interstate and intercity communications, but that 'last mile' in the switched network to the subscriber is still twisted copper pair, and will remain so for years.[17]

Time also felt that cable was the strongest contender for the text business for psychological reasons. This reasoning revolved around the fact that up until about 1980 people thought of TV as a way to receive three or four entertainment oriented television channels. However, with the advent of cable, dozens of channels focusing on a wide variety of subjects became the norm. All of a sudden, television became a more dominant and versatile part of the American home. Now there were even more compelling reasons for people to gravitate toward television for more hours of the day.

However, the questions remained. Was there enough interest to make teletext a viable business? When Time researched what others were doing, they discovered that virtually all of them had concentrated their financial resources on developing the technology and building networks. Very little time and money had been devoted to content. Furthermore, competing text services appeared to be a potpourri of generic news presented in a manner that mimicked electronic replication of existing print. With their reputation for journalistic excellence in magazines, Time had proved itself to be an excellent communications channel that knew how to profit from market segmentation. They now realized they had a competitive advantage in this area.

Magazines are aesthetically pleasing because of their judicious use of colors and graphics. Time was particularly convinced that the quality of graphics would play an important role in the success of their teletext service. However, they were careful to point out that they were not going to be carried away with fancy portraits or artwork, since overusing colors and gratuitious graphics, they felt, could have a negative effect on the viewer. While proponents of the World System firmly believed that visual quality was not important, Time was worried that their graphics would not be good enough in the long run. This is why, on Friday, 20 November 1981, they announced that they had adopted the North American Broadcast Teletext Standard for their national, satellite-distributed, full-channel teletext service. This made sense because the standard synthesized the best features of the Canadian Telidon and French Didon standards coupled with those of AT&T's Presentation Level Protocol.[18]

"The North American Broadcast Teletext Standard allows versatility and efficiency in the transmission and display of information," said Sean McCarthy. "We intend to take full advantage of its capabilities to offer the most innovative and attractive product to consumers."[19] Being at the leading edge of the technology was consistent with their overall view of competing in the marketplace.

This view seemed to be best captured in a motto sign which hung in Time's teletext newsroom. It stated:

It isn't important how many words go on a page, but how many ideas come off the page into the minds of the reader.[20]

In spite of the fact that decoders to support the NABTS standard were more expensive than those for the World System, Time felt the extra cost was well worth it. According to Pfister:

Even if the slight differential for higher-quality graphics remains, we believe the extra value will justify the cost. For example, you can still buy a black and white television set for less than color, but how many people choose their new television set based on price alone?[21]

Like other fledgling teletext services, Time was mainly stymied by the unavailability of low-priced decoders. Early discussions were carried on with several potential equipment suppliers with the aim of developing a one-hundred-dollar teletext decoding box. In fact, in late 1982, Time announced that it and Matsushita Electric Industrial Co. had agreed to codevelop consumer home terminals for Time's national teletext service. Time planned to have local cable companies purchase the boxes and distribute them to homes free of charge. The logistics would be similar to the arrangement for the decoder box for video distribution; the cable company owns the boxes but amortizes their cost through monthly customer fees which also cover program and service costs.[22]

To their credit, Time's instincts were good. Faced with the decoder dilemma and the realization that teletext would probably not negatively impact their core print business, they terminated their operations in 1983. Perhaps without knowing it, when Time exited teletext they took a great deal of the vitality out of the industry. If the highly respected Donaldson, Lufkin, & Jenrette study had crowned the wrong eventual big winner in the teletext sweepstakes, who then could carry the idea forward? When it came to the NABTS standard the answer was quite clear: broadcasters. However, the World Teletext Standard was still a powerhouse and clearly wasn't about to relent its early competitive edge to the NABTS standard. While the industry was badly shaken, it wasn't down for the count yet. Unfortunately, within a couple of years the broadcasters suffered the same fate as Time.

# 6

## Viewer Impressions

As we have seen, a number of surveys were conducted on the use and promise of teletext. However, since very few people ever saw teletext in the United States, any survey at all where viewers could express their impressions of the technology was worth analyzing. The author, who was highly involved with teletext, conducted a survey with graduate students enrolled in a college business telecommunications workshop in the summer of 1985. A little over half of those enrolled in the workshop were teachers and the remainder were employed in both the public and private sectors. After seeing a demonstration of CBS's "Extravision" service, only one of the nineteen graduate students felt teletext would never succeed and the longest time frame anticipated for its becoming a successful service was five years. Over 90 percent of them were wrong![1]

What is important to point out from this small survey is the likelihood that highly educated people may not necessarily have any more foresight on future market successes than anyone else. However, they may be somewhat more adept at identifying specific features and assessing their pros and cons.

Since the number of participants in the survey was small, their individual responses to open-ended questions have more relevance compared to statistical tests computed for various categories. With this in mind, a sampling of some of their comments on the advantages, disadvantages, and implications of teletext revealed the following:

### ADVANTAGES

- While viewing the graphics, the audio portion of the regular program can still be heard.

- It is very easy to learn how to operate the keypad which activates the TV to tune in Extravision.

- Teletext presents news stories with brevity and completeness.

- One can acquire news when one wants it.

- Information is constantly accessible twenty-four hours a day in the home via the television screen.

- The information is free and it is presented in a very unique manner.

In spite of the euphoria one might commonly experience after seeing something new and exciting for the first time, the students still astutely pointed out what they felt were some disadvantages of teletext.

## DISADVANTAGES

- The majority of the information is in written form and must be read, which denies the poor-sighted or illiterate.

- My eyes were strained looking at the screen for just a couple minutes. Different combinations of colors made the information hard to read. For example, red on green. Maybe it was the bright lighting in that particular room, but there is cause for concern.

- The print seems rather small and is similar to a computer readout, which can be a strain on the eyes.

- The choice of colors used also presents a strain for the viewer.

- With teletext you can only see information the broadcaster is sending.

- You must have good signals for reception.

- Its major drawback to future growth is that it is a one-way system.

- The only disadvantage of teletext is the switching from the programming to the system. Customers will not be able to watch their television program and teletext at the same time.

- Many people will not accept the idea that if they want teletext, they have to buy a Panasonic television.

- Waiting twenty seconds may be too long for some people.

- This is a strictly one-way communication—no feedback to the originator.

Anything new must compete with what exists. Its presence creates the potential for change. Here is a sampling of some insights into what the students felt the impact of teletext would be:

## IMPLICATIONS

- Teletext threatens classified revenues.

- The present high cost of decoders defers people's interest.

- Our lifestyles will be changed to the point where the media will give us only the news that it feels is important for us to hear/ see.

- A quarter will buy a newspaper and give the customer the ability to cut out various points of interest found on the hard copy to be used for future reference. It takes thirty to sixty minutes to learn about local and national news happenings by watching the "regular" news programming on television for no extra costs. For many people, it is part of their daily ritual to sit down, relax, and watch the newscast after eating dinner— something people look forward to as part of their "leisure" time. It's hard for me to understand why, at this point in time, anyone would want to spend five hundred to seven hundred dollars for a special TV receiver or decoder to obtain information at any time of the day or night. When the decoder is built into television sets for a nominal fee compared to the cost of buying the separate decoder at the present time, I can see people investing in the extra capability, but until that time, EX-TRAVISION will be used primarily by the person who has to depend on constant updates for business purposes and by people who always seem to be the first ones on the block to purchase the new product on the market, no matter what it is!

- Free!? Did I hear someone say that Extra-Vision was free? Pardon me, but according to the "Living Webster Encyclopedic Dictionary of the English Language," the definition of "free" is defined as ". . . without charge; gratuitously." Correct me if

I am wrong, but I don't believe there are any Panasonic dealers giving away "black box" decoders. It is my understanding that for the exchange of approximately seven hundred dollars American, I will in return receive a decoder. That is anything but "free."

- It will not take the place of the newspaper or the network news, but it will be a competitor. The advantages it will have over the newspaper and newscasts are the frequent updates and the selection of news for the viewer.

- It could definitely affect the lifestyle of many people by allowing them to plan other activities at times that they would normally be watching television. Perhaps more people would get in the habit of checking for the latest news and be more aware of what is going on in the world. For many it is difficult to be able to be in front of a television set at 6 P.M. or 11 P.M. to catch the latest news.

- Extravision will help and hurt various businesses in the future. First of all, in the future consumers might be able to access store specials and advertisements through EXTRAVISION. This will have a positive effect on the advertisers by increasing their sales. I think this will cause more competition in advertising in the future. As far as newspapers are concerned, I do not think they will be put out of business by teletext. . . . One section of the newspaper that I can see changing considerably is the classified section. In my opinion, all mediums of communication will be affected by teletext. The ones that accept the technology, research, and change will be the ones that survive and survive well.

- With teletext a business can provide the consumer with advertising in a more efficient way than newspaper and less costly fashion than television. In time people will refer to teletext to help with regular shopping chores. . . . It will also be the source of employment for many people.

- In the future teletext can do nothing but help business. Because we are becoming an information society, businesses can use this form of technology to their advantage. The cost will probably be less than advertising on network television and yet eventually reach the same people.

- One way for teletext to survive would be if there was resistance to the computer's use; people might use this service instead.

. . . At this time, I would see this service as a luxury item afforded by upper-income people. . . . People want information condensed, fast, and animated.

- As the call for more and better information develops in business and industry, so too are people beginning to demand information for personal use. By using teletext, people will become better informed, more productive personally and better consumers. . . . One distinct advantage at the present time is that you do not have to have cable or home computer availability to make use of teletext.

- It takes some time to overcome resistance to change and a fear of the unknown.

- It is easier for a lazy person to switch to EXTRAVISION and "read" the news than to get up, get dressed, and walk in the rain for the paper. . . . Being able to have the ability to aid hearing-impaired people is a great breakthrough.[2]

The previous year, a graduate student who completed an independent study with the author and later went on to become a lawyer, made some interesting and insightful observations while doing her research paper on teletext. Perhaps the most poignant of these came down to the matter of consumer awareness. In her words:

So what it boils down to is a case of prerogatives—at least in my case, since I already am "in the know" about all this trendy stuff. For the other people who said "WHAT? You're doing a paper on WHAT?", perhaps we should start by informing them of just what it is that they are passing up. Perhaps the biggest reason that teletext is not catching on is that people just don't know enough about it. Groveling through article after article really didn't help me to understand the actual systems. I guess I would have learned much more if I could have obtained information from the companies which actually broadcast teletext signals. . . . I don't exactly live in a poverty-stricken area, and I'm sure many of the people could easily afford a decoder. Unfortunately, as educated as my neighborhood masses might be, half of them haven't even heard of teletext or are aware that such services are currently available to them.[3]

## THE PRODUCT LIFE CYCLE

What these rather abbreviated academic opinions of teletext reveal is the fact that people are adept at identifying the pros and cons of new technologies but are slow to act, particularly when they feel they have insufficient information. This is nothing new. In the study of consumer behavior it has been repeatedly documented that consumers can be profiled according to their willingness to purchase new things. For the sake of convenience and clarity it might be helpful to cite these categories. As you can see in Figure 6.1, only two-and-a-half percent of the population can be termed innovators. They are quick to acquire new items and will often do so without having sufficient information prior to their acquisition. Following them in the growth of acceptance of a new product or service are the early adopters. Only 13.5 percent of the population falls into this category. Later, as the product/service becomes more widely known, the early majority, consisting of some 34 percent of the population, will purchase. Mind you, all of this unfolds over the course of months or years. Following the early majority are some 34 percent of the population that can be termed the late majority. They will follow suit only after the product/service has been available for a lengthy period of time and after the price has diminished considerably from its introduction. Finally, there are the laggards.

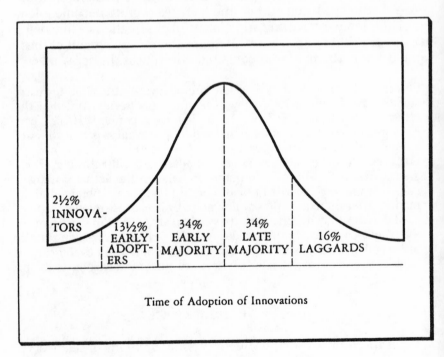

**Figure 6.1. The Product Life Cycle.**

They comprise about 16 percent of the population and acquire the product/service only after it has almost become obsolete. Needless to say, this group does very little for the success of the product/service, but in somewhat janitorial fashion they seem adept at picking up and disposing of the scraps after the product/service has run through its life cycle.

In the case of teletext, what is perplexing is the fact that there were just not enough innovators to drive teletext to the next level of consumer acceptance. The fact that teletext was found in only a few cities accounted for this impediment. Another barrier for the innovators was the dissimilarity of what they were being offered. It wasn't just plain vanilla teletext. Much rather it was different standards and competing delivery systems; two major points of confusion that proved to be too big an obstacle for very many innovators to overcome.

# 7

# Starting a Local Teletext Service

IN ORDER FOR BROADCAST TELETEXT TO BECOME A SUCCESSFUL NEWS dissemination service in the United States, all network affiliates would have had to participate. To do so would have required investing a substantial amount of money in teletext equipment. Perhaps this was one of the major reasons why ABC never pursued a teletext service. On the other hand, both CBS and NBC were always very enthusiastic and started national teletext services that led to several affiliates actually starting their own local teletext magazine.

Those network affiliates who decided to participate in teletext were able to receive national feeds from the network, create local stories, and integrate both in the vertical blanking interval during transmission of their regular programming. At a minimum, excluding personnel to operate all aspects of the system, this investment was, on average, at least $250,000 in 1984. However, as pointed out earlier, teletext decoders were brutally expensive and consequently almost totally absent from consumer homes during the period of time the networks generated a national feed. As a result, only the most risk-oriented affiliates were willing to take a chance.

## WIVB STARTS A LOCAL SERVICE

In the case of CBS, a couple of affiliates did invest in teletext and for a period of some two years provided both a national and local teletext service to their viewers. The first affiliate to provide this service was, as pointed out earlier, WBTV in Charlotte. On 25 July 1984, WIVB-TV, the CBS affiliate in Buffalo, N.Y., became the second station to begin a full teletext service (see Appendix 2). Like in Charlotte, the local service consisted of twenty pages. Local page creation using high resolution graphics was provided by Macrotel, Inc., a company which at the time specialized in videotex home banking, but was also heavily involved in electro-technologies research. Text information for the pages was gathered and edited by the WIVB-TV staff and transmitted on their VBI using Norpak equipment.[1] Two samples of WIVB local pages are illustrated in Exhibit 7.1.

## WIVB AND MACROTEL FORM AN ALLIANCE

WIVB-TV was, at the time, being managed by Les Arries. Under his leadership the station had repeatedly demonstrated a willingness and ability to be on the leading edge of the video industry. In addition to being a CBS affiliate they also owned a subsidiary that provided satellite uplink and downlink services. Their satellite dishes were located adjacent to the station and featured HBO as one of their main clients. This, of course, was at a time when the entertainment satellite business was just getting started, so through WIVB's pioneering efforts in satellite, Arries had gained a reputation for being a visionary in the video industry.

Macrotel was founded in August 1981, a joint venture between Empire of America bank and Leonard R. Graziplene, who became president of the venture. He had previously been heavily involved in video and was closely affiliated with Canada's Telidon protocol. The rationale upon which the company was founded was to provide the bank with a high quality videotex home banking service. Predating the founding of Macrotel, Arries and Graziplene had many discussions about the business opportunities presented by the fledgling videotex, teletext, satellite, and even wireless telecommunications technologies. By early 1983, WIVB had decided to take a leadership role in teletext and develop their own local pages. Following this decision, discussions and negotiations between the two companies intensified.

Because of its much greater revenue generation potential, Macrotel's primary interest was in videotex home banking. Obviously, the company did not have a broadcast television license, so broadcast teletext was quite marginal to their core business. What Macrotel did have that gave them an opportunity to generate revenue from teletext was two Norpak state-of-the-art page creation stations. In addition, they had two graphics artists on staff who were highly trained and talented in the use of these complicated systems. Realizing this, Macrotel proposed an arrangement with WIVB-TV that would enable them to reduce their original investment substantially by retaining Macrotel to create their daily local pages. Nothing came of this proposal for over a year, but being

Exhibit 7.1. WIVB-TV Sample Pages. Macrotel/WIBV-TV—Buffalo.

somewhat risk-averse, Arries was inclined to think that such an arrangement would make sense until it could be determined that broadcast teletext was going to be successful.

In early 1983, Macrotel had purchased a teletext decoder from Norpak and was able to access the CBC teletext service from Toronto. Since Buffalo cable television systems were carrying CBC programs into Buffalo, the teletext feed was received without typical over-the-air interference. Inasmuch as CBC was generating teletext signals in the VBI, it made it easy for Macrotel to view their service on a continuous basis.

After Arries had the opportunity to watch the CBC service at Macrotel, he brought Albert Crane, vice president in charge of teletext at CBS, in for a demonstration. Numerous pages were accessed for about forty-five minutes and the conversation centered around which protocol offered the best features. CBS at this time was using the French Antiope protocol, which had just begun featuring alpha-geometric graphics. The CBC teletext magazine featured the Canadian Telidon protocol, which may have had even superior graphics. The relative merits of the two systems were compared, especially with regard to costs and graphics quality. While there was no consensus as to which had the most to offer, all parties were sensitive to the fact that the North American Electronics Standards Committees were very close to adopting a standardized protocol for North America.

It primarily was based on the Canadian Telidon protocol and later supported by AT&T after having worked out some differences with the Canadian government. Unfortunately, AT&T was in the process of breaking up after losing a major lawsuit accusing it of being a monopoly in the telecommunications industry. Otherwise, it probably would never have taken so long to establish a teletext standard for the U.S. The acryonym for the protocol became NABTS (North American Broadcast Teletext Standard).

Because of the heavy expense of putting the technology in place to create and distribute teletext, it was very important to select the standard that would be legally adopted for the long-range future. Graziplene made the point that in the final analysis NABTS was going to be the winner and therefore the Canadian system, which was already in essence this standard, was the least risky to acquire.

Shortly thereafter, CBS made an announcement that they were going to use the NABTS standard for their service. Within a few months CBS was transmitting its Extravision teletext service from California. The only problem was, to whom? It now became critical for their affiliates to invest in two levels of technology; one, to access the national feed, and two, to add local pages and integrate both services into the VBI of their regular programming.

## MACROTEL INITIATIVES

Shortly after the original meeting involving Crane, Arries, and Graziplene, steps were taken to solidify the relationship among the three companies. Macrotel proposed five activities to WIVB which would make best use of the time of each company leading up to starting a local teletext service. Specifically, these were:

1. Establish and test the technical linkage between Macrotel's page creation systems and host computer/database, and WIVB-TV's broadcast hardware.

2. Determine the subject and content of general teletext pages with a local content.

3. Identify closed-user groups locally who would find it feasible to utilize teletext.

4. Because of your satellite uplink facilities it might also be wise at this time to interest a wide range of national clients to become involved in the early stages of teletext transmission. If you establish your expertise and leadership from the beginning, you will likely insure your playing a prominent role in teletext for many years to come.

5. Cost out all aspects of teletext transmission to ascertain the scope of revenue streams and potential for profit.[2]

The alliance was building and a sense of opportunity continued to grow. However, for a while things had to slow down because of a series of problems Extravision began to experience. Although CBS started incorporating a teletext magazine in their VBI on 4 April 1983, it was to be another year before an affiliate began to offer a local service. Indeed, conditions during the latter half of 1983 did not provide a great deal of enthusiasm or hope for the survival of CBS's Extravision service.

On 8 January 1984, Macrotel wrote a letter to Norpak, headquartered in Kanata, Ontario, Canada. Norpak was perhaps the leading Canadian manufacturer of Telidon hardware and was eager to sell its products in the United States. More specifically, they were interested in having CBS become one of their customers. As pointed out earlier, Macrotel specialized in videotex, but because of the compatibility of the videotex NAPLPS and teletext NABTS protocols, they were probing for a way to link videotex services to teletext, thereby increasing Macrotel's potential for revenue generation. In retrospect, it now appears that if teletext and videotex could have been merged into adjacent systems operating from a single source, the survival of both would have been enhanced. However, history tells us it is always difficult for two parties with compatible technologies to merge them for strength. Much rather there is often shortsightedness over a desire for independence and an overevaluation of the worth of one's own technology. In this case, an effort was made to create a system with more features, but even if it could have been achieved, there was still no reassurance that this approach would have fared any better than the independent routes pursued by videotex and teletext. A diagram of the system Macrotel was proposing, as well as seeking help from Norpak to configure, is in Figure 7.1.[3]

Later that month a meeting was arranged at CBS headquarters which was attended by Crane (CBS), Arries (WIVB), Graziplene (Macrotel), and Norton (Norpak). While all had their own agenda to solidify their respective positions in teletext, the nature of the meeting was primarily to address and help decide the system configuration WIVB would adopt for their service.

## Preparations Intensify

By early February 1984, CBS was again making significant headway and began to create added interest among its affiliates. Macrotel was now creating prototype pages for WIVB and checking their compatiblity with those being created and transmitted by Extravision from California. Within two weeks a number of steps were taken to strengthen the relationship between WIVB and Macrotel. First there was a technical problem to overcome. Macrotel had acquired a Norpak teletext decoder to access the signal from CBLT in Toronto, and were able to do so successfully. But when they tried to access the Extravision signal they were unsuccessful until they purchased a Panasonic tuner to attach to the Norpak decoder. For some reason the teletext signal emanating from Canada was different than CBS's. With teletext, it seemed nothing was compatible!

Beginning in 1983, Macrotel had been servicing a national database for Moneyplex centers (a financial services company) in vari-

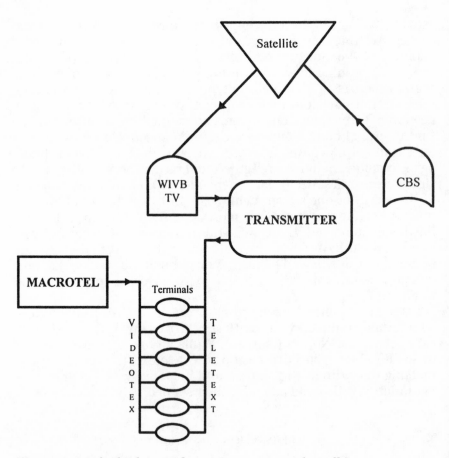

**Figure 7.1. A Hybrid Teletext/Videotext System. Macrotel—Buffalo.**

ous parts of the U.S. This required doing daily page updates and transmitting them to the banking centers where Moneyplex was located. Valuable experience had been gained, and this put Macrotel in a position to deal with a situation where they could just as skillfully do the page creation for WIVB's local teletext service.

Then there was also the matter of seizing the opportunity to solicit maximum press coverage from the initiation of a teletext service in Buffalo. It was agreed that the upcoming National Association of Broadcasters annual convention in Las Vegas would be an excellent forum for an announcement.

Finally, Graziplene shared a thought with the other parties that was designed to enlighten and strengthen the strategies that were building for the widespread acceptance and growth of teletext. In his words:

Perhaps the one last remaining mystery for all of us is the matter of deciding how to use teletext to insure a profit stream for all participants in our proposed project. The answer obviously lies in the area of applications; who would want to use this medium for what purposes. In many respects this is an even bigger challenge than unavailability of inexpensive decoders. While the issue at hand has all the outward appearances of the chicken and egg dilemma it is really up to a relatively few of us to demonstrate the brillance that will take this industry out of the dark ages and into the television screens of the viewing public. The bottom line is quite simple! How good are our ideas? A follow-up, brutally frank, meeting on all aspects of what is contemplated would do us all a world of good.[4]

Subsequent meetings were scheduled and a plan began to unfold. It now appeared that WIVB was finally committed to offering a local teletext service in Buffalo.

For its part Macrotel, in anticipation of becoming the page creator for the local magazine in July, approached its parent company, Empire of America bank, with a proposal for them to become the sponsor of one of the twenty daily pages WIVB would be offering in its local service. They declined to put up the money for the page, but later were given a free page as part of the compensation package Macrotel negotiated with WIVB to provide their page creation service. Interestingly, in spite of Macrotel being a subsidiary of the bank, there was still enough skepticism on their part for them to withhold even minimal support.

The local service amounted to twenty pages per day. The first tentative list of these pages can be found in Figure 7.2, and other details of WIVB's plans can be seen in Appendix 2. These pages were updated early each morning and then sent via telephone lines to WIVB-TV for insertion into their system. The primary focus was on news, sports, weather, traffic conditions, and investment updates. To stimulate interest in the service throughout Buffalo, WIVB placed TV sets equipped with decoders in ten public locations in the area to demonstrate Extravision (see brochure in Appendix 2). It was felt that interest would be enhanced because of updates on the upcoming Olympics scheduled to run from July 28 through August 13 in Los Angeles.

In these high-traffic areas it was assumed that people could become more familiar with the service by going through the procedures to access pages themselves. This went on for over a year and during this time the decoders became less expensive, but were still too costly for the average consumer to purchase. In spite of this huge obstacle, everyone associated with the local teletext service

TENTATIVE LIST OF WIVB-TV
"EXTRAVISION" PAGES

00 Master Buffalo Index

01 News 1

02 News 2

03 News 3

04 News 4

05 Weather Forecast

06 Extended Forecast

07 Stocks of Local Interest

08 O-T-C Stocks of Local Interest

09 Financial Tips

10 News 4 Sports Spotlight

11 News 4 Sports Briefs

12 Local Services Guide Index

13 Weekdays on WIVB-TV 4

14 4 Your Information

15 WIVB-TV/Key Bank Waterfront Festival

16 Entertainment Guide

17 Road Advisory

18 Call for Action Consumer Tip

19 Extravision System Locations

BUFFALO BROADCASTING CO., INC.    1077 ELMWOOD AVENUE    BUFFALO, NEW YORK 14207    PHONE  716 874-4410

**Figure 7.2. WIVB-TV Local Page Menu. WIVB-TV—Buffalo.**

was optimistic because several television manufacturers had announced that they were working on a teletext chip set that would soon be built into all new television sets and add only fifty dollars to the cost of the set. Inasmuch as the chip set would be integrated, much like UHF when it went from a black box on top of the television set to an integrated tuner, it was felt that this was the breakthrough needed to make teletext successful.

## THE QUEST FOR REVENUES

Aside from the problems associated with the delay in producing an inexpensive teletext decoder, there was also the continuous issue of deciding content for the service. Then too, for all the parties involved, there was the matter of revenue streams to consider. Obviously, for any business to succeed it must have sufficient profits.

By September, two months after the WIVB service began, discussions intensified on how revenue could be generated to support the teletext service locally. Several ideas were proposed in a letter from Macrotel to WIVB. Three of the most timely of these were:

1. A local database of some twenty pages is insufficient to derive a worthwhile revenue stream from when all investments are considered. In order for teletext to become profitable on the local level it will be necessary to establish a number of databases; thereby giving viewers better selections of high interest areas. . . . Of all the options considered, the most likely producers of revenue would come from these areas:
   A. Classified Ads—this could be done either in conjunction with a newspaper or else handled independently.
   B. Shoppers Services—among the participants that could become key components of this service are retailers, supermarkets, airlines, and real estate, to name but a few.
   C. News—this service is already being offered and does not require much more modification.

2. Although it may seem to run contrary to the objectives of broadcast teletext, the short-term use of cable makes sense. What is most important at this juncture in time is for there to be an adequate outlet to justify charging for the aforementioned services. Thousand dollar decoders are stifling the consumer market. Using a full channel on cable could reach a

sizable market and start the laborious process of increasing market awareness and consumer use. We have identified many of the details associated with such a joint venture, but in the final analysis, this is one of the very best options currently available.

3. Consider the potential of national closed-user groups. The ability to tie together widely dispersed interest groups via broadcast teletext has considerable appeal. Using an example from a widely read magazine, *Popular Science,* directing short features such as "Wordless Workshop" to highly segmented markets could prove to be highly successful. In this same context the use of teletext for a national real estate service with Gallery of Homes also has possibilities.[5]

During the early days of teletext there was no revenue stream, period! It was strictly missionary work and freebies. While it was quite clear from the beginning that revenues would be slim, this did not impede efforts to constantly think of ways to make involvement financially feasible. What it came down to from the beginning on both the local and national levels was a heavy concentration on two potential revenue sources; page creation and advertising. Both proved to be difficult tasks. For example, WIVB attempted to sell both in conjunction with the 1985 Auto Show & Exposition in Buffalo (see Appendix 2), but neither the Niagara Frontier Auto Dealers Association nor local ad agencies responded to their offer. It was now becoming quite clear that financial support for teletext both locally and nationally was not going to happen in sufficient volumes to support it, let alone make teletext profitable.

## SOME LAST GASP EFFORTS

In spite of the difficulties that were now plaguing teletext with a vengeance, pockets of enthusiasm and growth managed to keep teletext alive. For example, in October 1985, Smart TV in Tampa announced it would start a teletext service using the British World System Teletext standard. Although a private unaffiliated station, they estimated it would cost them about $15,000 a month, or $180,000 a year. They further estimated that if they could place 20,000 terminals they might net $3 million a year.[6] It is not surprising that even in the face of mounting odds, there is always someone who concludes they are the one who can succeed. Of course, being an optimist and having investors to back up one's vision help the cause immensely.

Sometimes news travels slow. In March 1986 the *Niagara Gazette* published a complimentary story on Extravision and WIVB's local magazine. While both were still operational at that time, it was life-support articles such as this that encouraged and provoked thoughts of a miracle happening.[7] Unfortunately, in spite of these reports of its vitality, teletext was on the brink of collapse.

## EXTRAVISION TERMINATED

By late 1986 most of the networks on both sides of the border had canceled their teletext service. CBS was one of the last to call it quits. When it did, WIVB and the other affiliates providing local pages also terminated their services. The end had come, but it did not stop there. Albert Crane was relieved of his duties at CBS at the time they terminated Extravision. Interestingly, he made an offer to CBS to acquire the rights to their vertical blanking interval. It was his intention to move ahead with teletext as a private venture.

## EXTRANET FORMED

In August 1986, Crane announced the formation of his new company which he called EXTRANET. The new corporation's primary purpose was to organize and operate a nationwide information distribution network using the television vertical blanking interval.[8] Not only did Crane negotiate with CBS, but also with other television networks and independent broadcasters as well as Wall Street brokerage firms and videotex companies. Even at this late stage Crane was optimistic that the long-awaited inexpensive decoders were going to be imminently available. This would then have permitted the realization of his plan to begin to structure and implement the commercial applications of data broadcasting that would offer businesses nationwide the opportunity for low-cost access to traditional information sources.[9]

Crane felt that Extranet would gain market acceptance for three reasons:

1. Competitive pressure has placed a premium on rapid (electronic) delivery of decision making information.

2. Information that is currently printed and mailed can be delivered immediately by Extranet, can be viewed by a subscriber—or can be directed to a personal computer for analysis, filing, or later viewing.

3. The Extranet method of information distribution costs only a fraction of traditional electronic methods such as telephone or telex, and is more flexible than newer methods such as Direct Broadcast Satellite (DBS).[10]

Extranet's initial marketing and service thrust was directed towards providers of current market information, publishers of industry newsletters, and publishers of professional and reference services. Crane marketed his new venture at most of the major telecommunications conferences in the U.S. and at the same time attempted to raise money from interested investors. While he openly stated that it would take five to ten years for his service to catch on in the U.S., he also stated that "telecommunications is a thirteen billion dollar business and I am just looking for $30 to $40 million of it."[11] However, in the end CBS and the other networks refused to deal with him. With this last gasp effort having failed, broadcast teletext passed from the scene in the U.S. by early 1987.

## SUMMARY

There is a lesson to be learned from the Extranet initiative. Individuals who have management responsibilities in ventures being pursued by large companies can easily conclude that they, not the company, are the key factor in moving an idea to fruition. However, the truth is that the large company, not any one individual within the company, is almost always the electromagnet to attract and amplify on opportunities. An individual who leaves the company with the intention of pursuing the idea in their own venture will find the going rough, because they no longer have the power and resources of the large company to draw from. In this case, if the networks couldn't make it work, why would anyone conclude that an individual with a new venture could make it work?

In retrospect, the efforts expended to make broadcast teletext a viable business could hardly be faulted given the maze of impediments to be overcome. Success can be achieved when participants work well together. Certainly, for example, the alliance between Macrotel and WIVB made good business sense. All during the relationship, risks were shared and each genuinely tried to continuously improve the service and identify revenue streams. As planned from the beginning, Buffalo was in the lead in promoting broadcast teletext in the U.S., but when it all ended in late 1986, the fact remained that the industry died and money was lost.

# 8

## The Failure of Teletext

TELETEXT PEAKED IN THE UNITED STATES IN 1984. ACCORDING TO Albert Crane of CBS, teletext was expected to be a good business for CBS affiliates by 1985, with the break-even point coming before the end of the decade. Barbara Watson of NBC expected to show a profit in teletext by 1987, with 10 percent of American homes taking the service by 1990. Both missed their predictions, but the question forever remained, why?

Videoprint predicted that because teletext could be updated instantly it would compete heavily with evening papers for post workday readership. Furthermore, their report contended that teletext's around-the-clock availability would free viewers from the programming schedule to which nightly television news programs bound them. Finally, they concluded that the immediacy and user-friendliness of teletext would also make that medium a formidable competitor. But, when they also projected a teletext market of about five hundred million dollars by 1994, with 75 to 80 percent of those revenues coming from advertising, they too missed their mark by a mile.[1]

By early 1985 the media began to express doubts about teletext's future. There were reports that teletext had not lived up to early expectations, and recent moves could be interpreted either as the first signs of its demise or, hopefully, the beginning of a more realistic outlook on the part of the teletext industry. But because there were as yet no really cost-effective decoders, which research suggested should cost less than twenty-five dollars to be acceptable, teletext revealed its weakness. The two main questions for the future centered around whether the technology would arrive in time, or whether the standards battle between interest groups really killed teletext. 1985 was the year when answers dominated.

There was now widespread consensus that teletext had not lived up to early expectations. And, in spite of supposed progress having recently been made, the press began to interpret these assumed gains with considerable skepticism.[2] In this case the press was accurate in its appraisal, for by the end of the year it was evident that the industry had failed in the United States.

Why teletext never caught on in North America like it did in Europe remains a mystery. Some observers felt that the introduction and growing popularity of the VCR during the early to mid-1980s was the main reason. Still others cited the steep cost of decoders needed to display teletext services on their television screen. There was also the possibility that teletext would have infuriated program sponsors by allowing viewers to zap commercials in favor of quick news updates. Another damaging factor was the long delay in setting standards for the medium. The NABTS standard adopted for North America was, strangely, different than the one followed in Europe. In the final analysis there were just too many damaging reasons, and these accounted for the woes and quick demise of both broadcast and cable teletext.

## Closed-User Groups

When teletext was first introduced it was felt that it was best suited to be a mass-audience service. However, as teletext began to fail in this respect an effort was made to direct teletext to specific needs and groups. A highly defined market segment became known as a closed-user group. A number of these were identified and an effort was made to make these profit centers. Two of the most notable were the investment industry and agriculture.

### The Investment Industry

Of all the projected uses for teletext, none sparked as much enthusiasm and expectation of financial gain than the decision on the part of investment companies such as E. F. Hutton to utilize teletext for transmission of securities quotes. Here finally was a market sector overflowing with cash and with a clientele that could benefit directly from teletext on a daily basis.

While at first this may have seemed a good idea, it only could work if investors had no alternatives to turn to in order to get quotes. However, this was not the case. Several television programs such as CNN provided this service as part of their regular video coverage of news. An investor could also purchase or lease a telephone device to give them quotes. Both of these options were far less expensive then a teletext decoder. Furthermore, since most investors didn't even know about teletext there was little likelihood the idea would easily catch on.

By far the greatest impediment to teletext being used for invest-ment purposes was its close relative, videotex. Rather than being content to have information flow to customers on demand, brokers were more interested in earning commissions through trades; a process made much easier if they utilized videotex rather than tele-text. That is why E. F. Hutton and other brokers quickly considered teletext but in the end were much more infatuated with the pros-pects of what videotex could do for their business.[3]

Agriculture

One of the most interesting pilot studies that considered the use of teletext by agriculture and agribusiness was conducted by the University of Wisconsin Extension (UWEX). They called their demonstration "Infotext."[4]

During the last half of 1985, UWEX, using a World System Tele-text version in their Infotext database, conducted a demonstration of teletext for those in agriculture. Their plan was to provide a ser-vice over the five network and two affiliate stations of the Wisconsin Educational Television Network. Decoders were placed in exten-sion service offices and in ninety-four farms and agribusinesses that had volunteered to participate in a nine-month field evalua-tion. Ameritext, the U.S. representative for the world system tele-text format, agreed to provide UWEX with a Jasmin teletext editing station and a one-hundred page teletext insertion system. Green-dale Electronics, another British company, agreed to provide a number of leased and free-loan set-top decoders. Zenith Electron-ics, the only American manufacturer of world system teletext de-coders, donated a number of television sets equipped with decoders. For its part, UWEX agreed to conduct research to deter-mine the potential audience for an agricultural teletext service, keep an accounting of costs, determine the technical quality of the service, and report on the reliability of the decoders.[5]

A panel of sixty-eight farmer-evaluators and agribusiness users rated teletext during this period, and over two-thirds of them indi-cated that they would be interested in purchasing teletext decoders in the two-hundred-dollar range when they became available. The most useful reports, according to the participants in the study, were the state radar, wind speed/direction, and temperature maps pro-vided by Infotext. Interestingly, and perhaps a sign of things to come, 50 percent of the PC owners preferred to have Infotext teletext pages delivered to their computers rather than their television sets, and were willing to pay an additional one-hundred-dollar equipment fee for this convenience.[6]

Like most of the other studies that preceded it, this one focused on teletext as a tool to help a vested interest group do a better job for its constituency. Taking the lead in proposing and enacting change insulates the knowledgeable reputation of the research agency and can serve as sort of an advertisement for the good the agency is doing. Unfortunately, in agriculture, as well as other industries, essential information can typically be gleaned from many existing sources. Teletext, although making information available in a different format, never did eclipse or add to the database of information already available. For this reason there was no compulsion for those in agriculture to convert to teletext as their preferred channel of vital information.

This was an interesting study and there was much to be learned from what was one of the late research projects on the use of teletext. During the course of the study, Greendale Electronics stopped production of decoders for the North American market.[7] This study, like most analyses of teletext, demonstrated an interest, but the market could never develop to take full advantage of what might have become a highly useful service for farmers.[8]

## Some Major Impediments to Adoption

As we have already seen, there were many impediments of different magnitude that haunted teletext throughout its developmental stage. Although we have considered many of them in some detail, it may be advantageous to amplify our analysis of some of them.

### AT&T Breakup

In the year 1982 the telecommunications industry experienced a serious change of direction. This was precipitated by the Justice Department's settlement with AT&T which resulted in the virtual breakup of the company. While they were required to divest themselves of such significant entities as the Bell Telephone System and Western Electric, AT&T retained the right to be in the information business. With this obstacle removed, AT&T was positioned to move into a business strongly linked to a delivery system of which they owned 95 percent. While later legislation was introduced in an attempt to get AT&T out of the information business, one fact that was to play a significant role in the development of teletext soon became apparent. The rates for using the telephone system would soon go up substantially because of deregulatory moves and the ending of subsidization of local service by long distance revenues. Consequently, the most attractive alternative for disseminating electronic news was broadcast teletext.

One can only speculate, but it is highly likely that if AT&T had not been forced to downsize at the very time when teletext, videotex, and related technologies were being introduced, the outcome would have been different. In spite of its court battles, AT&T continued to have an influence on electronic technologies. With regard to standards, AT&T was a prime supporter of the North American standard (NABTS). What AT&T failed to do, however, is get directly and highly involved in the evolution of the electronic technologies of the time. Unlike in Europe, where telephone companies were the major players in the development and implementation of teletext/videotex, AT&T was conspicuously weak in her involvement. Indeed, the print media sustained a much higher profile during the early 1980s. The truth is, teletext looked like something that should have come from AT&T; not the television industry. This point should not be discounted in any analysis of why teletext failed in the United States.

Advertising

One of the major attractions of teletext was its assumed appeal to advertisers. The nature of teletext led some observers to conclude that it would be cheaper for advertisers to reach their desired audience with teletext because there would be no paper, printing, or handling costs. Furthermore, advertisers could benefit from the fact that they would be reaching potential customers at the precise moment when they develop an interest and want information to satisfy it. Finally, the cost of teletext advertising was felt to be extremely low given the fact that a television station could send out up to twenty-five pages of teletext print for display on a viewer's TV screen in one-tenth of a second. It would take between six and eight minutes of broadcast time to deliver a comparable advertising message on commercial TV.

The cost of teletext advertising was projected to be only a fraction of the rates that would have to be charged for ads printed in newspapers, and viewers were expected to prefer teletext because it would enable them to easily and conveniently compare specifications, ingredients, features, and prices of different brands and competing products. With such assumed severe electronic competition, it's no wonder the print media felt compelled to become so highly involved in teletext.

## Standards and Delivery Conflicts

Since teletext was resident in the vertical blanking interval it was possible to deliver the signal either through a cable or through the air. However, the choice of delivery brought with it specific restrictions. Delivery through the air limited the number of pages to approximately two-hundred, but if cable were utilized instead the number of pages of information that could be offered escalated to five-thousand. One of the questions from the beginning concerned whether the best service consisted of more or fewer pages. Needless to say, broadcast and cable television stations saw themselves as competitors from the beginning, thereby adding to the confusion confronting would-be viewers.

Of all the problems encountered, the most difficult to resolve was perhaps the respective strengths of the competing standards. The intensity of this issue fragmented the future prospects of teletext. If the market could talk it might have been with two loud voices. On the one hand it was widely believed by broadcasters in the U.S. that teletext would be an advertiser-supported medium, and therefore optimum graphics were a must. They also felt that this approach would be more aesthetically pleasing to viewers. The opposing view championed the European experience, where teletext gained its earliest success, and seemed to negate the importance of graphics, both to advertisers and viewers, in favor of lower hardware costs.

The two competing standards were also quite different because of the cost associated with each. Nowhere was this more prominently in evidence than in the huge difference in the cost of decoders. At the height of its popularity, world standard decoders were priced in the neighborhood of three hundred dollars, whereas NABTS decoders never reached less than two hundred dollars. Ideally, a chip-set built into all new television sets might have diminished this difference substantially enough to bring universal acceptance of the industry.

## REGULATORY ISSUES

With the advent of teletext, U.S. policymakers were confronted with a series of tough questions. For example, how was the spectrum to be allocated? What technical standards should be adopted? How should teletext be licensed? What regulations, if any, should govern the editorial content sent over teletext systems? Most of the burden of answering these questions rested with the Federal Communications Commission (FCC).

In 1980 the FCC was already being barraged with requests from the private sector on this matter. CBS, for example, filed a petition with the commission, urging it to reserve lines 10 through 16 of the vertical blanking interval for teletext. As an added pressure, until the commission acted, teletext was relegated to remain in the experimental state, since it could interfere with existing broadcast operations of other new services such as closed-captioning for the hearing impaired.

CBS also requested the FCC to adopt the French Antiope protocol as the technical standard for all U.S. teletext systems. Not to be outdone, a short time later the British petitioned to have their protocol become the standard, and last but not least, Times-Mirror and Time, Inc. pressed hard to have the Canadian Telidon version of teletext become the North American standard. And, somewhat later, AT&T unveiled their own Presentation Level Protocol (PLP) and made a strong case for its adoption. Clearly, the FCC was now under the gun and the legal profession started to line up in anticipation of having a fertile ground of suits to settle for years to come.

It got even more difficult for the FCC very quickly. If broadcasters were to be required to make their teletext facilities available to a newspaper publisher or anyone else, the 1934 Communications Act would have to be amended. The commission could have solved this matter quickly if they had decided to license teletext services separately from standard broadcast operations. But then if they did decide to issue new licenses exclusively for teletext services, they would have had to develop criteria for selecting among competing applicants.

As a final tangle to complicate the growth of teletext, there were several other content regulations to consider. The most important of these were the Fairness Doctrine, the equal opportunities for political candidates issue, the reasonable access issue, the personal attacks and political editorials issue, and the obscenity, indecency and profanity issue. How did each apply to teletext? Well, again, until these questions could be answered teletext was held hostage, unable to move ahead unimpeded.[9]

To be more specific, even before it gained much attention the question was being asked, "Where does teletext fit in the American regulatory structure?" To help answer this question the obvious starting point was the 1934 Communications Act. It states that broadcasters and common carriers are subject to different sets of obligations. Broadcasters have certain statutory obligations with respect to the content of communications delivered to the public, but no obligation to provide service to third parties as a common carrier. Common carriers, on the other hand, are generally viewed as having no responsibility for the content of communications transmitted, but do have an obligation to offer service to third parties upon request and are subject to public review of rates, terms, and conditions of service.

One major problem arose when the FCC refused to classify cable television services as either broadcasting or common carriage. Another appeared on the horizon when it was realized that there were two distinct types of teletext services would have to be dealt with in the future. First were the narrowband teletext services that offered captioning, advertising, and other program material closely related to broadcast television programs.

Second, there were freestanding narrowband or broadband teletext services, with the latter occupying a full UHF, cable, or Multi-Point Distribution System (MDS) television channel. Still another problem revolved around the fact that under provisions of Title III of the Communications Act, there was nothing to prevent broadcasters from contracting out the management of their teletext service. Just as long as there was reasonable monitoring of the teletext service by a licensee, the FCC could conclude that a licensee should not be held strictly accountable for the misdeeds of a third-party program service supplier. And to further add to the dilemma, there was still the question of how legal requirements would affect the marketing of teletext services embedded in broadcast signals carried on cable systems.

As teletext appeared to grow so too did the number of lawsuits, especially between cable and broadcast stations. For example, WGN channel 9 in Chicago at one time sought to enjoin a satellite common carrier (United Video) from stripping out WGN's teletext signal in its VBI prior to distribution to cable systems around the country. What provoked this outrage was United Video's decision to offer cable systems a teletext signal provided by Dow Jones in place of WGN's.

Unfortunately, guidelines for future business arrangements were being decided in litigation before the longer term business implications of different outcomes could be fully comprehended. Within this upsetting milieu the FCC hesitated to adopt any standard that seemed likely to favor one system over another or to ban any one system. It is very likely that had broadcast teletext survived, its growth would have been an agonizing process made all the worse by legal constraints that, from the beginning, created a ready platform for litigation.[10]

## TELETEXT'S INHERENT LIMITATIONS

Teletext was a complex medium that included more than graphics and reading. As we have seen, a whole series of problems were present from the beginning, many of which involved inconvenience and presentation nuisances. Among these, some of the most damaging either not already mentioned or insufficiently considered include: high costs, lack of a critical mass, slow access time, limitation of screen size, incongruent choices of colors, positioning limitations, and age of television receivers. Each had its share of problems. We shall consider each of these areas separately to learn more about how they contributed to the demise of teletext.

### Cost

The fact that broadcast teletext is resident in the vertical blanking interval of the video signal is a convenience that made it seem all the more likely that it would become a popular consumer service. From the broadcasters' perspective there was relatively little extra expense in delivering a teletext service. It was like giving viewers an extra channel for virtually no additional cost. The fact that the service was free to viewers was an incentive, but the cost of the NABTS (the standard eventually favored by the major service providers) decoder proved to be a major disincentive. In 1983 a NABTS teletext decoder was priced at over three thousand dollars. By 1986 the price had dipped to somewhere in the range of nine hundred dollars, and two years later it was available for approximately two hundred dollars. Arguably, even at two hundred dollars there was little perception of a bargain for accessing news that could be received (usually later) in a variety of less costly ways. Of course, to a relatively few viewers, there may have been value received from accessing desired news on demand.

### Critical Mass

Originally, it was thought that decoders the size of chip-sets would be available to build into the TV set. This thinking closely paralleled the procedure followed in the late 1950s when the UHF signal, soon after being activated by the FCC, could only be accessed by a separate tuner placed on top of the television set. Later, as we know, set manufacturers integrated the UHF tuner into all new television sets. This resulted in an extra cost, but it was small and inconspicuously passed on to the consumer. Nobody complained. Perhaps if a similar tack had been pursued earlier for broadcast teletext, and added only a nominal fifty dollars to the price of a new television set, we might now be looking at several successful national broadcast teletext services. Unfortunately, the old chicken and egg fable prevailed and we now have neither the service nor mass-produced television sets with integrated teletext chips.

## Access Time

Broadcast teletext was brutally slow! Because the entire package of some two-hundred pages of news was scrolled, it was possible at times for it to take as long as twenty seconds for a specific page of requested information to appear on the screen. For most viewers, this was too long a wait. Understandably, people don't like to wait in lines and are generally impatient when having to deal with anything that frustrates their sense of immediate gratification. Perhaps something could have been done to dispel the viewers' impatience when they were compelled to wait so long at times to get the information they requested.

Clearly, a short access time was critical. Since this was not possible, a more practical solution would have been to provide some type of feedback that, at the very minimum, simulated something happening. For example, when making a phone call, you don't expect to immediately talk to the person the moment you finish dialing his or her number. Very often the person calling will tolerate many rings before hanging up and trying again later to reach their party. Perhaps teletext could have borrowed this technique and provided some mechanism that signaled the viewer that the page desired was being called up. Indeed, introduction of a sound similar to the ring of a telephone might have been sufficient to at least partially dispel the impatience resulting from frequent aggravating delays.

## Screen Size

The amount of space within which a screen of teletext information could be displayed was very limited. It was tempting but unwise to pack too many words on a screen. Doing so made it highly difficult to read and destroyed the aesthetics of the information. One of the biggest problems with the average television set in the United States at the time was its rather small size; 17 to 19 inches. The average size screen in Europe, on the other hand, was 26 inches. By some accounts it was necessary to limit the number of words on a teletext page to between 75 and 100. Even this number was probably too many. There was already the problem of getting people comfortable with reading rather than watching a television screen, so the only solution would have been to use much fewer but more powerful words.

## Word Engineering

Teletext could never successfully be treated like print, even though it was in some ways "word engineered" illustrated news. To get the most out of the medium it might have been wise to follow the methods and techniques of cartoonists. Perhaps no writers can put as much meaning into a compact space to make a point as the authors of comic strips. They are masters at using a powerful blend of words and illustrations to convey messages that can readily be understood by almost everyone.

The advertising industry has repeatedly proved that the ability to be noticed and responded to is heavily influenced by the judicious use of a few well-chosen words. For example, if someone wanted to get your undivided attention they might choose to swear at you. One or two short four-letter words, carefully selected, would not only gain your immediate attention, but at the same time clearly convey the hostile feelings of the sender. There is some speculation that teletext would have been an excellent medium to introduce a new powerful information medium driven by word symbolism, graphics, and animation.

## Choice of Colors

Everyone is influenced by color. It is probably safe to say that most people have a favorite color. However, color is a very complex thing. In all cultures it is closely related to psychology. Depending on the choice of colors, an object or event may be interpreted in a narrowly defined unique way. It is uncanny how certain colors have come to represent certain things in all cultures. In the United States, for example, from a marketing point of view, we commonly associate the colors orange and black with Halloween; red and green with Christmas; red with Valentine's Day; and red, white, and blue with the Fourth of July. Black is the color of sorrow and the color white is the favorite of brides-to-be. There is ample evidence that colors strongly influence our personal psychology.

In addition to the underlying meaning assigned to colors, there is also the matter of color coordination to consider. Some colors blend well with others, but some clash. Knowing this, the wrong choice of color combinations in mass communications would undoubtedly prove to be an impediment. For example, combinations of red and blue are particularly difficult to read. In broadcast teletext this lesson was often unheeded and one can only speculate what this error might have done to undercut the appeal of teletext.

## Symbolism

If there had been more time to treat teletext as a communications art form, the new technology might have achieved a major breakthrough in reducing communications to simplicity. Scholars have long contemplated the prospect of reducing the vast differences in the diverse languages of the world with universal symbols. We already see indications of this, for example, in the use of a caricature of a man on the door of a restroom for men rather than the words "men's room." Along the highway the signs put up by repair crews will often contain no words, but instead show the caricature of a person using a shovel. Someday, less to describe more will be the norm; exactly what teletext gave us, even if it was only a brief glimpse.

## Old Television Sets

When teletext was first introduced in the United States the main object of concern was the high cost of a decoder so users could access a service. Little mention was made of the fact that all older television sets were not equipped to access teletext even if a decoder was plugged in. Consumers stood a chance with a recently purchased set, but to be safe one needed an RGB monitor. These were even more expensive than decoders. Therefore, in the early 1980s consumers might have been compelled to spend as much as five thousand dollars to access a service that was being advertised as free. This was perhaps the steepest barrier which kept teletext from becoming a highly prized consumer service.

## The Survival of Teletext Remnants

Even after the networks had shut down broadcast teletext by late 1986, there was still an impression that teletext was a viable service. In one report, for example, which appeared in February 1987, it was pointed out that teletext had grown in one year from a single national service to four different services. These included:

ELECTRA—an electronic newspaper that offered instant access to news, sports, weather, business, and general family information. Approximately fifty pages of information were available. It was available on Galaxy 1 (channel 18), on Satcom 3R (channel 6), and on cable with Superstation WTBS or the Tempo Network.

TEMPO TEXT—an electronic stock market quote service, continuously updated throughout the day with closing prices displayed overnight until trading resumed the next business day. Stock quotes were delayed from actual trading by fifteen minutes. The service was available on exactly all the same sources as ELECTRA.

INFOTEXT—primarily an agribusiness information service, with current grain, livestock, and dairy reports from Wisconsin and the U.S. Department of Agriculture. In addition, commodity prices from the Chicago Board of Trade and the Chicago Mercantile Exchange were updated after a fifteen-minute delay. Other information included AP news and sports stories. This service was offered on Galaxy 1 (channel 22) and on cable with the Discovery Channel.

DATAVIZION—a family fun and information magazine with unique news stories, trivia, games, horoscopes, a satellite TV program guide, and more. It included seventy-five pages of information updated once per day. It too was available on exactly the same sources as INFOTEXT.

Based primarily on the British protocol, receiver/decoders were claimed to be readily available from such sources as American Teletext, Dick Smith Electronics, and Zenith Electronics (Digital System 3 TV sets with built-in teletext decoders).[11] Even with the availability of these vestiges of teletext it wasn't enough to make teletext, in the final analysis, a successful industry. These services were limited in scope and never would have achieved the broad market penetration of a network teletext service. Companies always feel they are filling a need, but there is always the question whether they will be permitted to fill the identified need.

## FILLING A NEED

Given the importance of communications, it is puzzling to determine why this promising technology failed in what is generally acknowledged to be the most information oriented society in the world. In spite of its problems, logic dictates that teletext should have (could have) been a resounding success. To emphasize this point, think for a moment how difficult it is to access information on an as-needed or desired basis. Where, for example, could you turn to get a summary of all the most recent world news headlines in the next minute? Or, for that matter, what if you wanted to know how your favorite sports team was doing in an important game? Realistically, there really aren't too many options available to this day.

In spite of all its impediments, teletext should have survived until technical problems were alleviated and a critical mass of customers generated. After all, this is no different than the growth pattern expected by almost all innovations.

Availability of news only gets worse on the weekend. Now that most newspapers specialize in morning editions there is a tremendous time lag in keeping pace with news events occurring between Friday and Monday. Radio fills part of the gap, but you really never know for sure when news is going to be presented or how comprehensive the report will be. Teletext would have been an ideal medium to consistenlty keep updated news flowing to television viewers. But teletext really never had a fair chance to prove what it could do. As a result the U.S. took a step backward in the area of information technology.

## DOES TELETEXT HAVE A FUTURE?

What are the prospects of teletext being available in the U.S. in the next few years? History provides us with some interesting guidelines. If one subscribes to classic theory in attempting to time the mass acceptance of technology, William F. Ogburn's "cultural lag" theory might have relevance for the future of teletext. He contended that most new technologies usually take approximately one generation (twenty years) from the time they are first introduced to the time they will be adopted and utilized by the general population.[12] If this is true we can look for teletext to start gaining acceptance sometime around the turn of the century, based on the fact that the British invented the technology during the 1970s.

On the other hand, alternate technologies, especially fiber optics and satellite distribution of television programs, may have already totally eclipsed the need for teletext. And with HDTV soon to become the new television standard in the United States, the higher resolution and 3 x 5 aspect ratio will certainly alter the configuration of the current vertical blanking interval where teletext resides.

The availability of teletext, or something similar, would almost certainly have a major effect on lifestyles. Where today we are compelled to wait for the media to provide us with desired information at random intervals or in rigid blocks of time, teletext would change all that by offering immediacy of information. While teletext can't go into great detail describing each news story (television and newspapers can do that), abundant facts can be made to flow quickly. The full ramifications of what this could mean for lifestyles is too early to determine. However, at the very least it could contribute to freeing up time and putting people more in control of their activities.

In this fast-paced stressful world of ours, there never seems to be enough time to get everything done we would like to accomplish. Only those technologies which are perceived to offer high value, personal satisfaction, and have psychological "sizzle" will likely succeed in the future to gain access to minds that, through years of experience, have learned to erect row upon row of barriers. Teletext is a medium restricted to the use of very powerful words accompanied by equally powerful symbols and graphics.

As things now stand, we may never know or feel the impact of teletext. The key to what happens to the technology is probably still in the hands of the major television networks. During recent years CBS, ABC, and NBC have all lost market share, and will likely lose more in the years ahead. With the proliferation of more networks on cable, broadcast has given way to narrowcast. This may mean the networks will be financially stressed to resurrect teletext. However, it is interesting to note that some of the most popular network programs are news oriented. Couldn't teletext be a powerful adjunct to these news features?

In addition, television viewing is under increasing pressure from alternative forms of leisure time experience. For example, more people are spending increasing amounts of time on their computer; both doing tasks and going online. The phenomenal growth of the

Internet, especially the World Wide Web, is testament to a major shift in lifestyles.

Teletext is a centralized technology. It should make it in an environment that largely nurtures narrowcast over broadcast and individual choice over bureaucratic control. On the other hand, quiet television can't make it. There can be no substitute for high drama in news dissemination. On-site video has already eclipsed newspapers as the favored source of news in the U.S. and nothing seems destined to change this.

If anything, teletext will probably go down in the annals of communications history as one of the technologies that facilitated the transition from analog to digital video. Or perhaps, it was the appetizer we needed to lend impetus to the popularity of the Internet. It gave us cause to contemplate our preferences for news delivery formats. However, with attention spans diminishing, only the most avid reader would be willing to tolerate the slowness and rather bland ambience of teletext. On the other hand, somewhat coincidentally, many of the problems encountered by teletext have also been experienced by the Internet. The most major of these have been lengthy delays in accessing information, or worse yet, being unable to get on the Internet whenever desired. The aesthetic and information configuration of web pages is confronted with the same set of issues once faced by teletext. And, a ton of regulatory issues are about to surface.

Suffice it to say that technologies are very finicky. Teletext may one day return with a vengeance, but for now, rather than lament the failure of a technology from the past, we are much more inclined to assess the steady stream of new technologies. It is somewhat ironic, but in the end we will probably settle on the scantiest pieces of news to satisfy our curiosity; the very quality that once upon a time was the hallmark of broadcast teletext.

# Appendix I

## THE WORLD TELETEXT STANDARD

—World System Teletext—Now
—Keyfax National Teletext Magazine
—World System Teletext—Zenith

**WORLD
SYSTEM
TELETEXT**

**NOW**

World System Teletext Brochure. Keycom—Schaumberg, Ill.

## What Is Teletext

Teletext is the transmission of text and graphics supplementing regular television service. Viewers can, whenever they want, select from a variety of pages (screens) which include national and *local* program schedules, news, weather, sports, and advertising, as well as community information.

Teletext is transmitted as hidden data pulses along the vertical blanking interval (VBI) of broadcast or cable TV; it can also be transmitted via all the lines of the TV screen. The viewer simply uses a hand held remote control to command his teletext decoder-equipped TV set to display the particular screens required.

As an advertising medium, teletext can exploit its advantage through being constantly updated and being available to viewers at all times when normal TV is broadcast. Teletext can readily carry full screen advertisements or commercial product reminders, at the top or bottom or elsewhere, of screens carrying normal teletext information.

## How Teletext Is Produced... Editorial Center

Editors can produce pages from standard information sources, such as program schedules, weather reports, sports scores, newsroom copy or newswires. Newswires can, in fact, be interfaced directly to the teletext computer. The diagram below shows how simple it is for a main editorial center to be integrated with normal broadcast TV services.

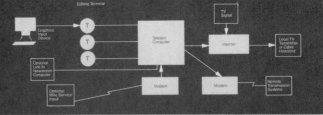

## How Teletext Is Produced... Local Center

Local TV stations or cable operators can relay a teletext service from a national, regional or associated editorial center, adding local information such as program listings, community events, weather, advertising and on-screen flashes, including emergency information, weather, traffic and travel news. In such applications, a single editing terminal will normally be adequate. In addition, a data bridging device may be used to move a teletext service from a satellite or associated TV station signal to the local VBI.

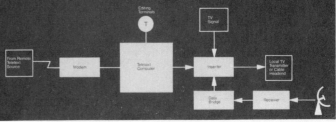

**World System Teletext Brochure. Keycom—Schaumberg, Ill.**

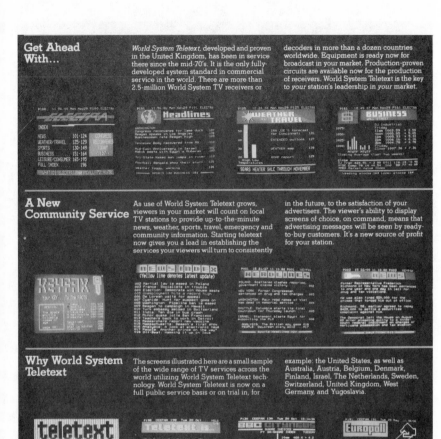

**Get Ahead With...**

*World System Teletext,* developed and proven in the United Kingdom, has been in service there since the mid-70's. It is the only fully-developed system standard in commercial service in the world. There are more than 2.5-million World System TV receivers or decoders in more than a dozen countries worldwide. Equipment is ready now for broadcast in your market. Production-proven circuits are available now for the production of receivers. World System Teletext is the key to *your* station's leadership in *your* market.

**A New Community Service**

As use of World System Teletext grows, viewers in your market will count on local TV stations to provide up-to-the-minute news, weather, sports, travel, emergency and community information. Starting teletext now gives you a lead in establishing the services your viewers will turn to consistently in the future, to the satisfaction of your advertisers. The viewer's ability to display screens of choice, on command, means that advertising messages will be seen by ready-to-buy customers. It's a new source of profit for your station.

**Why World System Teletext**

The screens illustrated here are a small sample of the wide range of TV services across the world utilizing World System Teletext technology. World System Teletext is now on a full public service basis or on trial in, for example: the United States, as well as Australia, Austria, Belgium, Denmark, Finland, Israel, The Netherlands, Sweden, Switzerland, United Kingdom, West Germany, and Yugoslavia.

World System Teletext Brochure. Keycom—Schaumberg, Ill.

Keyfax National Teletext Magazine Brochure. Keycom—Schaumberg, Ill.

# KEYCOM
## Electronic Publishing

## What KEYCOM Is About

On April 22, 1982 Centel Corp., Honeywell, Inc. and Field Enterprises, Inc. announced the formation of a joint venture to provide videotex and teletext services to the business and consumer markets, a venture which its planners say will "bring a wealth of information and communications capabilities into each person's home and business."

### KEYTRAN

The joint venture, KEYCOM Electronic Publishing, will develop, market and operate a complete videotex service called KEYTRAN, which will allow home and business users to retrieve information and transact business on a TV screen. The videotex service will begin operation in the Chicago area in mid-1983, with plans for subsequent expansion to other markets.

Using KEYTRAN, the videotex service, the consumer can access thousands of pages of written and graphic information, perform two-way interactive functions such as user-to-user communications, and transact business including home banking and home shopping.

The basic KEYTRAN service, including a videotex terminal, is expected to cost less than $25 per month. This relatively low price is made possible by a specially-designed Honeywell videotex terminal, which is compatible with the proposed U.S. standard announced by AT&T.

Large Honeywell mainframe computers will be used at the system's center to ensure quick response time to user requests regardless of how many users are connected to the system. State of the art technology, ease of operation and top quality service are bywords for KEYTRAN service.

### KEYFAX™

KEYCOM's teletext service, KEYFAX, is a one-way system by which users can select up-to-the-minute news, sports, business, weather and leisure information. Currently KEYCOM produces and distributes KEYFAX and NITE-OWL, a late-night teletext-type program, via broadcast over Channel 32 (WFLD-TV) in Chicago. In addition, KEYCOM and Satellite Syndicated Systems have reached agreement to market a national version of KEYFAX to cable operators which serve more than 20 million homes throughout the country.

KEYFAX customers use a hand-held numeric keypad, and select "pages" of information for display on decoder-equipped televisions. The expected monthly rate for KEYFAX service is $10.

The KEYFAX system utilizes multiple computers and is based on proven British teletext technology. KEYCOM is the leader in bringing teletext service to the consumer and it is intended for KEYFAX to provide the highest quality and most desirable services available.

KEYCOM Electronic Publishing  Schaumburg Corporate Center  1501 Woodfield Rd.  Suite 110 West  Schaumburg, IL 60195    (312) 490-3200

Keycom Electronic Publishing Brochure. Keycom—Schaumberg, Ill.

# KEYCOM
## Electronic Publishing

## How KEYTRAN and KEYFAX™ Work

"The KEYCOM videotex and teletext systems will not only provide instant access to a wealth of information and services, they will do so in a way that's both simple and reasonably priced," said Robert W. Nichols, president and chief executive officer of KEYCOM Electronic Publishing.

### KEYTRAN (VIDEOTEX)

Videotex customers will have a small terminal that connects to a television set and to a telephone line. The terminal includes a portable, alpha-numeric keyboard with all of the function keys necessary.

No installer is needed. The terminal can be easily connected to the antenna leads of a TV set, and plugged into an electric outlet and a telephone jack.

To operate, the user presses a button on the keyboard that instructs the terminal to make a local telephone call to the host computer. The computer then determines the user's identity as a customer of KEYTRAN, and a message, such as "Welcome to KEYTRAN," appears on the TV screen.

The system is designed for easy use and no prior knowledge of computers is necessary. The customer simply answers questions which appear on the screen. There's also an option to override this simple—but deliberate—selection process by entering "key word headings," which more specifically identify the subject material the user wishes to view.

The terminal, which contains memory/storage capability, allows users to "download" games and other information, so as not to tie-up the telephone line.

Individuals with personal computers will be able to program their units to access KEYTRAN videotex service as well. These customers will pay for KEYTRAN service, but not for a terminal.

### KEYFAX (TELETEXT)

Teletext customers will have a television set with a small decoder and a remote control numeric keypad. The decoder will enable the user to gain access to the teletext system. With the hand-held keypad, the user can find out what information is available on the KEYFAX system and then select the desired "pages."

Using the decoder, the user switches from broadcast transmission to teletext transmission. Using the keypad, the viewer then calls up a "table of contents," which provides page numbers for topics available on KEYFAX.

After selecting the topic, the request is entered by pressing the appropriate page number on the keypad. The selected page appears on the television screen within seconds and remains until the viewer requests a different page.

KEYCOM Electronic Publishing  Schaumburg Corporate Center  1501 Woodfield Rd.  Suite 110 West  Schaumburg, IL 60195     (312) 490-3200

Keycom Electronic Publishing Brochure. Keycom—Schaumberg, Ill.

# KEYFAX™

## NEWS OPERATION

A novel publication, called KEY-FAX, rides the airwaves. Transmitted on an unused portion of the television signal, this electronic magazine is America's first national teletext publication.

"Teletext" is the name for the revolutionary new system that puts words on your television screen. KEYFAX - your key to the facts - provides the latest news, information, and entertainment in the form of "pages" that are displayed on your television.

A television set, teletext decoder, and a handset are all you need to access KEYFAX. Press the number of your page selection on the handset and, within seconds, you'll receive the desired information.

Teletext offers you three advantages:

**MEDIACY:** It provides the *latest information* 24 hours a day.
**CONVENIENCE:** You read this information *when you want to*.
**CHOICE:** Choose what you want to read from over 100 pages of information available.

## MAGAZINE PRODUCTION

The KEYFAX magazine and its constantly-updated news, sports, business, weather and traffic information is produced by a team of specially-trained journalists.

Some journalists on the team spend their shift speed reading stories which flood into the newsroom on computerized links. Breaking news and special interest stories are selected, then displayed, on input computer screens run by American ATEX computers. There the editing process begins.

The story is first fitted into the teletext format of 40 characters by 24 rows. This process can often be expedited in a single step through the use of programmable keys. These keys save time important in this instant medium.

The story is next reduced to 75 words - the ideal capacity of our teletext pages. This is not easy. The journalist must present all significant facts and at the same time create a report that is attractive to read. A writing style unique to the

medium has evolved. Copy is presented as a clear, concise summary of the original text.

Once a story has been edited on the ATEX input terminal, it is sent to the output side of the newsroom. Here a different computer takes over. The output computer is called a CONTEXT computer system. It is a British system developed by the LOGICA company in London.

This system drives the teletext editing terminals and a device called a digitizer. These are used to refine and prepare pages for final transmission through the creation and insertion of graphics and color

## GRAPHICS

Graphics are constructed manually by skilled journalists at the output terminals or through use of the digitizer.

The digitizer is a device that can quickly translate, or "digitize," a well-defined photograph or drawing into a teletext computer graphic. The image can be touched up with a light pen.

Graphics serve two functions on teletext. The first is to create an attractive page. Graphics highlight the text, adding visual interest and appeal.

Secondly, graphics aid in identification for advertising. KEYFAX is a commercial medium supported by advertising revenue. This advertising takes a unique, informational form because the consumer controls selection of product information.

The KEYFAX magazine also utilizes graphics to differentiate sections of the magazine. Each section deals with a different topic identified by a page header which literally and graphically describes content.

**Keycom Electronic Publishing Brochure. Keycom—Schaumberg, Ill.**

# KEYFAX™ National Teletext Magazine

## FEATURING:

**NEWS**
Fast, accurate and factual reports on international, national and regional stories are continuously updated 24-hours-a-day.

**SPORTS**
In-progress reports on all major games plus results, statistics and standings from across the country are included with previews, personality profiles and commentary.

**BUSINESS**
From stock quotes to home finance tips, the business section follows the action on Wall Street, the NYSE, AMEX, COMEX, dollar and gold markets. Analyses of fast-breaking financial stories round out the section.

**WEATHER**
Complete regional weather reports are dramatically presented through colorful graphics and detailed forecasts.

**SPECIAL FEATURES**
Regional features on the people and events that make America thrive are top priority in the Nationwide Section. Viewers send greetings and make their opinions known in the special features section that also includes horoscopes, recipes, quizzes, reviews and much more.

Keyfax National Teletext Magazine Brochure. Keycom—Schaumberg, Ill.

## Introducing KEYFAX

No doubt you've heard of the "information revolution." People have talked about it for years now. But few have actually been able to take advantage of the remarkable advancements in the technology of information delivery.

Until now.

Because now, KEYFAX National Teletext Magazine is ready to come into your home.

### "But what's a teletext magazine?"

Essentially, KEYFAX is very much like a printed magazine or newspaper. With KEYFAX, however, the news is "delivered" over the cable system to your television set. The technology has been in use in numerous countries, such as Great Britain, for a number of years.

### "Do I need special equipment?"

When you subscribe to KEYFAX, you receive a decoder, which is attached to your TV, and a remote control handset. The handset, by the way, converts your standard TV into a completely remote-controlled television. Once attached, this equipment allows you to select interesting and colorful pages of information. You simply look at the index and enter the number of the page you wish to read. That page is then displayed on

your screen and stays there until you request another page. More than 100 pages of information are always available.

### "But why a teletext magazine?"

The greatest advantage to KEYFAX is speed. Because the magazine is continuously updated, rewritten and "delivered" to you instantaneously, you are able to retrieve the most current news, sports and business information available. Another benefit of the teletext magazine is convenience. You pick the information you want to read when you want to read it.

### "So KEYFAX is basically a news service?"

Basically, yes. But there's more. Like conventional publications, the KEYFAX Magazine contains numerous special categories and feature sections. For instance, if you're looking for something different for dinner, take a look at your KEYFAX Recipe Page. Wondering how your stars line up? Consult your KEYFAX Horoscope. Cartoons. Games. Quizzes. If it's of interest to you, you're likely to find it on KEYFAX.

### "Is it expensive?"

No, not really. You receive the KEYFAX decoder, handset, and full 24-hour access to the service for something less than 70¢ a day.

**Keyfax National Teletext Magazine Brochure. Keycom—Schaumberg, Ill.**

Zenith World System Teletext Brochure. Zenith—Glenview, Ill.

# ZENITH PRESENTS
## DIGITAL SYSTEM 3 WITH "WORLD SYSTEM" TELETEXT...

Television customers who will buy new Zenith Digital System 3 receivers can, for the first time, enjoy World System Teletext displays through decoders built right into the set. This major advance in home electronics coming soon, will give viewers the benefit of instant information on a variety of subjects, and offer TV broadcasters the opportunity to expand communication services to the public with increased revenue to themselves.

Teletext is a service developed by the BBC to transmit text information along with a regular television signal by utilizing the unused lines in the vertical blanking interval of a station's transmission. Generally, there are 3-5 lines available to transmit extra information such as teletext.

When you offer Teletext, you are offering the viewer a free additional programming service, aside from your regular programming. A viewer can get variety by simply switching to Teletext, while staying on your channel. The audio portion of your programming continues as he views Text, so he is able to monitor regular programming.

## ...AND A HOME TELETEXT PRINTER

Along with World System Teletext, Zenith will introduce an accessory printer that can be attached to a new Zenith Digital System 3 set. The printer will enable viewers to print out Teletext pages on the spot. Think what this can mean to you, the broadcaster, and to your advertisers when you are able to convey printed information directly into the customer's home. Now is the time for stations to take a hard look at World System Teletext and the unique opportunities available.

**Zenith World System Teletext Brochure. Zenith—Glenview, Ill.**

## WHAT'S GREAT ABOUT WORLD SYSTEM TELETEXT...

World System Teletext is the hardware/software system developed by the British ten years ago and used successfully in international markets ever since. It's main advantages over the NABTS system are:
- relatively inexpensive equipment purchase and operating costs;
- ease of operation;
- adaptability to existing broadcasting equipment;
- reliable and rugged performance; and
- quick displays of selected pages.

Another major advantage of World System Teletext is that it can coexist with certain Closed Caption systems. While the Teletext signal exists in one field of the VBI, Closed Captions can be transmitted concurrently in another field. This would allow you to simultaneously transmit Closed Caption and Teletext, and offer *both* to your audience.

## WORLD SYSTEM TELETEXT OFFERS "PAGE" AFTER "PAGE"...

The Teletext system continuously broadcasts a series of coded signals from which any page, or single screen of information can be selected via the TV's remote control. The cycling rate, which is the repeat rate of the pages, is dependent on the number of "pages" being transmitted. Teletext is capable of offering the customer a choice of as many as 300 different pages of information at any one time. The customer decides what page he wants to view by selecting a category and page number from the Teletext index or "menu," then entering that page number through the remote control provided with a Zenith Digital System 3 set which controls TV functions and Teletext selections. This remote control operates all functions of Teletext. It even has a unique Text update feature that can be programmed to let the viewer know when a particular story

has been updated. When new information is added to the story, an update warning will flash in a corner of the screen.

## ...AFTER "PAGE" OF INFORMATION.

Teletext lets the broadcaster transmit just about any information suitable for home viewing. This can range from printed matter, such as news headlines or sport scores to specialized material for advertising requirements. Here are some suggestions:
- News and sports headlines direct from the wire services, updated every few minutes.
- Financial market information as fast as it's received in the newsroom.
- Listings of local events, such as movies or concerts with curtain times and ticket prices.
- Weather forecasts and bulletins updated more frequently than normally broadcast on news programs.
- Up to the minute traffic reports.
- Special entertainment or educational material for children, such as quizzes with answers that can be revealed by pressing a button on the remote control.
- Public service announcements that can be repeated continuously, such as descriptions of missing children, availability of medical assistance, police and fire emergency information.
- Advertising messages, such as special sales and product descriptions tailored to a specific audience. For example, cooking

recipes, with the ingredients on sale at a local market.
- Restaurant specials with a map showing location.
- Classified ads and Real Estate listings.
- Special local news and business information.

## AND FOR YOU, THE BROADCASTER...

**Zenith World System Teletext Brochure. Zenith—Glenview, Ill.**

# YOU CAN GENERATE REVENUE FROM TELETEXT...

You, the broadcaster, can profit from Zenith's decision to produce TV sets with Teletext capability. You can convince your advertisers that ingenious marketing ideas can be applied to Teletext with virtually guaranteed results! For example, a local sporting goods store can sponsor a Teletext page of special interest such as a page of baseball scores. That page naturally attracts baseball fans. The advertiser's message is displayed as a footnote to the score page. Or in the case of multiple pages where scrolling is necessary, such as displaying a long list of batting averages or league standings, an entire page could be devoted to a sponsor's message.

# ...AND YOUR ADVERTISERS CAN MAKE MONEY!

This "sponsor's message" page then hits the Teletext viewer with a "hard sell" list of items on sale, and may include a coupon that is good for a discount at the sporting goods store. Think of the benefit of the Zenith Teletext printer. The viewer at home presses the "print" button on his remote control. The discount coupon is printed, and the consumer redeems the coupon on his next trip to the store... putting more money in the hands of the store owner, your customer! This also means the station can provide the advertiser with absolute proof of reach. The bottom line is: advertisers can buy Teletext space assured that their message is being targeted to the most likely prospective customer group and receive proof of its effectiveness.

**Zenith World System Teletext Brochure. Zenith—Glenview, Ill.**

## THERE IS A BIG POTENTIAL MARKET FOR TELETEXT

In addition to sales of Zenith Digital System 3 sets with built-in Teletext capability, there are already 4.5 million Zenith TV sets in homes that can use the existing Zenith World System Teletext Decoder. These are 4.5 million potential customers ready now for Teletext broadcasts. We believe today's TV customers are more feature conscious than ever, and Teletext is destined to be the next most-wanted feature.

## THE COST IS REASONABLE… TO BUY AND OPERATE WORLD SYSTEM TELETEXT EQUIPMENT

Basic equipment includes a computer, editing terminal and monitor, and graphics display unit. Annual operating expenses should be limited to engineering and editing requirements, and related marketing services. An example of one station's actual World System Teletext costs appears on a later page in this booklet. Basically, a reasonable rate structure combined with good selling techniques should result in earnings far in excess of purchase and operating costs.

## NOW IS THE TIME TO GET STARTED WITH TELETEXT!

We are convinced that Teletext is destined to be a winner for both viewers and broadcasters. An aggressive TV station can get a real jump on the competition by recognizing the audience drawing power of Teletext services and the potential for substantial increases in revenue. That's precisely why Zenith decided to produce Digital System 3 Television sets with built-in World System Teletext decoders.

Here's an opportunity for you to get a headstart on the competition by being the first to offer the latest in TV technology… Teletext. As a Teletext pioneer, your station can enhance its image as an aggressive leader who wants to be at the forefront of your field.

## HERE'S WHAT SOME CURRENT TELETEXT BROADCASTERS HAVE TO SAY ABOUT THIS VALUABLE SERVICE…

**Zenith World System Teletext Brochure. Zenith—Glenview, Ill.**

The Taft Broadcasting Company in Cincinnati is one of Teletext's pioneers, currently running an aggressive teletext program called Electra. Taft's senior management have taken an active part in Electra and are some of its most ardent enthusiasts. The following are excerpts from speeches they have given around the country to encourage the development of Teletext.

## TAFT EXECUTIVES ENDORSE TELETEXT...

"...A study by a major management consulting firm predicts that within ten years, home information systems such as Teletext will generate $10 billion in advertising revenue. That figure represents 10% of all advertising."

"...With equipment and staff in place, what can a local station offer viewers? Something new and different and unlike any other source of information. Teletext provides the fastest information service available to the public. The only way you can get current information any faster is by having an AP or UPI printer in your home. That's not hype; that's fact. As fast as the writers can rewrite the wire service stories or the local news stories on the teletext terminals, the viewer can get the information."

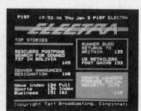

**Zenith World System Teletext Brochure. Zenith—Glenview, Ill.**

## ADVERTISERS CAN TARGET SPECIFIC AUDIENCES

"...Among the many values of Electra to advertisers is its "selective use" characteristic that

allows advertisers to target specific audiences with defined interests. When an advertiser sponsors a sports page, for instance, he'll be reaching viewers so interested in sports that they actively call-up that particular page of information. There's no "waste" of the advertising message. It can be targeted to key prospects. The speed, simplicity and low cost of production make advertising on Teletext unique from any other medium. Advertising copy can be produced and on air in minutes. Changes can be made instantly. Ads can be full or partial pages."

## TELETEXT IS FAST, VERSATILE AND INEXPENSIVE

"...Teletext isn't a substitute for other media. It's a fast, versatile, and inexpensive complement to other media. We expect a lot of our regular advertisers to tag their normal television commercials with references to more detailed copy they place on Teletext. There's even more potential for classified and real estate advertising on Teletext. And since Teletext can be broadcast 24 hours a day, there are hundreds of opportunities for advertisers to display their messages."

"...Whenever possible, Teletext copy includes direct references to upcoming station newscasts and news personalities. As Electra grows in popularity, that kind of cross-promotion will make it a valuable extension of the station."

## TELETEXT IS UP-TO-THE-MINUTE

"...Most teletext research to date indicates that the value of the service to the viewer isn't just the availability of the information, but the immediacy and up-to-the-minute nature of it. Without question, usage increases and declines directly in proportion to the frequency of updating. The more frequently the information is updated the more viewers will access teletext.

"...Unless there are attractive and interesting teletext services on the air and they are promoted effectively, there is no reason to expect consumer interest. Broadcasters have to provide the programming to stimulate the market and create an audience. The situation is similar to the development of color television. Just as color television strengthened and increased our viewership, so will teletext. With the competition from other new technologies increasing, broadcasters need to offer more viewing incentives."

## CAPITAL COST OF "ELECTRA" WAS ONLY $185,000

"...The total capital expenditures involved in putting Electra on the air were approximately $485,000. That price included the host computer with custom software, three editing terminals and monitors, a graphics display unit and miscellaneous equipment such as modems, multiplexers, and office equipment. We purchased a top-of-the-line system with capabilities over and above the minimum requirements for a typical station teletext operation. We probably will use smaller systems at our other stations when we initiate teletext in those markets."

## ANNUAL OPERATING COSTS ARE $135,000

"...The total annual operating expenses are approximately $135,000. That includes salaries for managing editor and three staff writers, half the salary of an engineer shared with the station, news wire service, advertising, marketing and research, telephones and miscellaneous office supplies. As we did in designing the hardware system, we approached the operations side of our teletext experiment in a very ambitious way. For many stations, a two-person staff probably would be sufficient. Most of the local information is provided by the WKRC-TV News Staff."

**Taft executives say "...TELETEXT IS AN EXCITING NEWS MEDIUM...IT HAS TREMENDOUS LOCAL POTENTIAL THAT SMART EXECUTIVES WON'T IGNORE."**

**Zenith World System Teletext Brochure. Zenith—Glenview, Ill.**

# UNIVERSITY OF WISCONSIN'S INFOTEXT HELPS FARMERS

Pilot studies conducted by the University of Wisconsin Extension (UWEX) indicate that teletext may serve to meet the information needs of another homogeneous, but geographically dispersed audience: agriculture and agribusiness.

In 1985, UWEX conducted the nation's first demonstration of teletext for the delivery of agribusiness information. The Infotext teletext magazine was broadcast statewide over the five network and two affiliate stations of the Wisconsin Educational TV Network. Decoders were placed in public locations, UW-Cooperative Extension Service Offices, and one hundred farms and agribusinesses for one month field evaluations.

## HOW INFOTEXT BEGAN...

The UWEX developed a prototype of Infotext for demonstration at two major 1984 farm technology events: The World Dairy Exposition and Wisconsin Farm Progress Days. Hundreds of visitors stopped at the Infotext display, and many expressed interest in having the service made available to them. At each event, a brief questionnaire was distributed surveying topics of interest that Wisconsin farmers and agribusinesses would like to see on teletext. The Infotext magazine was based on the results of 250 completed surveys.

**Zenith World System Teletext Brochure. Zenith—Glenview, Ill.**

In addition, the questionnaires solicited participants for a field test. Respondents were offered the possibility of receiving a teletext decoder for a one month evaluation period. An evaluation panel was selected, consisting of 24 agribusinesses and 70 farmers in 37 Wisconsin

counties. They represented the target audience for the Infotext service: the state's middle to large grain, livestock, and soybean producers.

Agribusinesses that used decoders in their lobbies or waiting areas responded very favorably to Infotext. Because Infotext was heavily accessed by their customers, many said it would be a valuable addition to their waiting rooms or lobbies. They indicated a willingness to purchase teletext equipment.

The response of the home/farm evaluation panel was overwhelmingly positive. Participants perceived Infotext as a provider of valuable market and agricultural news, and a superior weather information service. They rated teletext superior to radio, television, newsletters, farm magazines, and telephone-based on-line services in terms of convenience, timeliness, and content. Over 2/3 said they had a definite interest in purchasing teletext equipment.

Infotext viewing ranged across the entire magazine and throughout the day. The evaluators accessed Infotext from early morning to mid evening, with over 50% of the viewing occuring from 9 a.m. to 3 p.m. Weather information pages had the greatest viewership (98%) and some of the highest rated screens in the magazine.

All trial participants were users of the mass media, and subscribed to specialized farm magazines and newsletters. Many of the agribusinesses subscribed to commercial on-line videotext services. Yet, all judged

Infotext as equal or superior to these media in terms of convenience, timeliness, and content.

A full report on Infotext may be obtained by contacting:
Steve Vedro, Associate Director,
Telecommunication Laboratory
University of Wisconsin-Extension
230 Lowell Hall
Madison, Wisconsin 53706
Phone: 608/262-6135

## ZENITH SUPPORT OF INFOTEXT

Zenith established its own support of Infotext by supplying Zenith television sets and Zenith teletext decoders to nearly 100 dealers throughout Wisconsin. These were provided for an ongoing in-store demonstration of Infotext.

**Zenith World System Teletext Brochure. Zenith—Glenview, Ill.**

# KTTV INAUGURATES TELETEXT FOR '84 SUMMER OLYMPICS AND BEYOND

KTTV in Los Angeles initiated their Teletext service, Metrotext, shortly before the start of the 1984 Summer Olympics. During the games, Metrotext was operated as an information service, providing scores, athlete profiles, information about Los Angeles and more.

The Olympic Metrotext service was seen in 800,000 cable homes as a full screen rolling page magazine and on 100 specially equipped Zenith receivers around the city in high traffic areas such as shopping malls, hotel lobbies, tourist attractions, and restaurants. A questionnaire with a mailback card was left at each of these locations. Over 700 of them were returned, with 85% of those responding showing a real interest in the service.

## METROTEXT GROWS ON CABLE TV

The initial success and positive feedback to Metrotext resulting from the Olympic coverage encouraged KTTV to develop the Metrotext program. Metrotext is now seen on a dozen cable systems as a full screen service of the cable company's basic package. This translates to a current viewership of approximately 500,000 cable homes throughout the KTTV coverage area.

Metrotext feeds 50 pages of material, 24 hours a day to the cable companies in the form of rolling pages. Each page stays on the screen for about 20 seconds. (Time length can be varied on a page basis.) Information on the Metrotext magazine includes:
- Local news stories
- Sports, entertainment, and weather information
- Skiing conditions
- Quizzes
- Personal classified ads
- Nightly cable schedules
- Community events
- Human interest stories

Metrotext is actively seeking additional cable companies with interest in a teletext channel. In most cases, those with free channels will provide use of a channel at no cost.

Reaction to Metrotext has been overwhelmingly positive. Cable companies are pleased with the service, as they have received positive feedback from subscribers. Viewers are interested and ongoing survey results are encouraging. Press reaction shows a high level of interest and has helped to increase sponsor interest.

**Zenith World System Teletext Brochure. Zenith—Glenview, Ill.**

# Appendix II

CBS TELEVISION NETWORK
PRESS INFORMATION
51 WEST 52 STREET
NEW YORK, N.Y. 10019

## FACT SHEET
## TELETEXT AND EXTRAVISION™

### BROADCAST TELETEXT

Broadcast teletext is a generic term for one-way information systems that transmit letters, numbers, and graphics over the verticle blanking interval (VBI is the "black bar" that you see when the verticle hold on your television set is misadjusted). The information sent via this unused portion of the television signal is simultaneously broadcast with the network program, and is seen as a full page of information on the television screen.

### NABTS

NABTS (North American Broadcast Teletext Specification) is a broadcast standard that is an extension of the French Antiope and Canadian Telidon systems. This standard allows the broadcaster greater flexibility in design with its high graphic capability, myriad of colors, and ability to create limited animation on the screen. The NABTS standard is supported by CBS, NBC, and AT&T.

### DECODERS

Consumer decoders are currently being manufactured by Panasonic and Quasar, and are available through the set manufacturers. Sony has announced the production of its consumer model NABTS teletext decoder and will be selling it later in 1984.

### CLOSED-CAPTIONING

Closed-captioning for deaf and hearing-impaired viewers in the NABTS standard provide a wide range of caption services. For example, NABTS captions can be transmitted in different colors to differentiate between speakers; captioning may appear in different locations on the screen, as well as at multiple speeds; they can be transmitted in multiple languages; and many graphic symbols are used to indicate sounds off screen, such as a doorbell or telephone ringing.

### EXTRAVISION

The teletext service of the CBS Television Network is called EXTRAVISION.

EXTRAVISION is an "on-demand" teletext service; there are no scrolling pages, only directly accessed information that you want to see, when you want to see it.

### SERVICE CONTENT

EXTRAVISION is a 100-page electronic magazine that is constantly updated 13 hours a day, and accessible 24 hours a day in your home via the television screen. The comprehensive national and local service provides news, weather, sports, and feature information. Your local CBS affiliate station may provide pages of EXTRAVISION content unique to its viewing audience.

**CBS EXTRAVISION Brochure. CBS—New York.**

# Some Questions And Answers About

# EXTRAVISION™

**What is Extravision?**

Extravision is the name of the broadcast teletext service of the CBS Television Network.

**How does it work?**

Information is entered by the editorial staff into the vertical blanking interval, and is transmitted along with the regular broadcast signal.

To receive Extravision, the viewer's set must be equipped with a special decoder. This decoder captures the signal and reconstructs it into letters, numbers and graphics on the screen. By entering different codes on the key pad, the viewer can select "pages" from the Extravision magazine for display on his own television screen. It is "user friendly."

**What is the NABTS standard?**

NABTS, for North American Broadcast Teletext Specification, is a standard for creating and transmitting high resolution text and graphics. It is an extension of the French Antiope and Canadian Telidon systems. It has been adopted and is currently in use by CBS, NBC, and AT&T.

**Where can viewers buy a teletext decoder?**

Consumer NABTS decoders are available for purchase at select television dealerships within broadcast areas where local and national Extravision is being transmitted.

NABTS decoders are available from Panasonic and Quasar. Sony, also, has announced its support of the NABTS standard. For decoder purchase, contact a manufacturer's dealership near you.

**In addition to news, weather and traffic, what other local information can be offered?**

TV listings, movie schedules, book, movie, drama, dance and music reviews, ticket information, airline schedules, shopping information, consumer tips, recipes, puzzles and games to entertain the younger set...and much, much more.

**Is Extravision's closed-captioning for the hearing-impaired community compatible with other systems?**

CBS will expand its existing captioning efforts on a transitional basis and broadcast closed-captioned programs in dual-mode for the deaf and hearing-impaired viewers. (Both Line 21 and NABTS standards.)

The additional capabilities of Extravision will allow hearing-impaired viewers to see CBS's shows with several improvements like captioning in different lettering heights and colors to distinguish between actors, multiple speeds and placement of text, and the use of various languages.

**Are any Affiliates presently offering Extravision?**

In early 1984, 85% of the CBS affiliated stations (reaching 71 million homes) are passing the network signal. With a decoder, Extravision's national programming can be available to you for use right now. WBTV, in Charlotte, NC, was the first CBS Affiliate to go on line with both network and local pages.

**Is there any additional investment an Affiliate will have to make to participate in Extravision?**

There is no additional investment if an Affiliate chooses to pass along only what programming is being transmitted nationally. However, to create local pages an Affiliate would need additional equipment.

**Do any surveys indicate that Extravision improves regular program viewing?**

In the 1982 Los Angeles research study, 30% of the new viewers tuned into teletext from no prior TV viewing. Of the 30%, 60% went on to watch television programming, thereby increasing the overall viewing audience.

**Why get involved in broadcast teletext now?**

For an Affiliate, early participation leads to a competitive edge over other electronic delivery systems that are rapidly becoming a part of our everyday lives.

An advertiser can benefit from early experience with direct electronic marketing such as dealer tags, preparing him and his clients for the way of the future.

For a network, the plus is additional public services, such as closed-captioning for the hearing-impaired viewers, and offering further support to affiliate bodies, positioning stations in the new electronic media.

**If I have any questions about CBS's Extravision, who can I contact?**

Albert H. Crane, III, Vice President
Extravision Service
Betsy Illium, Public Relations
Extravision Service

CBS Television Network
51 West 52 Street
New York, New York 10019
©1984 CBS INC.

**CBS EXTRAVISION Brochure. CBS—New York.**

CBS TELEVISION NETWORK
PRESS INFORMATION
51 WEST 52 STREET
NEW YORK, N.Y. 10019

## TELETEXT AND EXTRAVISION ™ BACKGROUND

| | |
|---|---|
| 1972 | Teletext systems first demonstrated in the United Kingdom by the British Broadcasting Corporation (CEEFAX) and the Independent Broadcasting Authority (ORACLE). |
| 1974 | Antiope teletext system demonstrated in France by Telediffusion de France. |
| 1979 | Interest in teletext development in the United States formalized by the creation of an industry study group on teletext - formed at the request of CBS - under the auspices of the Electronic Industry Association's Broadcast Television Systems Committee. |
| January 1979 | CBS begins comprehensive experimental program to determine technical specifications for a teletext system that would be compatible with the U.S. broadcast environment. |
| March 23, 1979 | CBS begins on-air test of teletext signals at CBS Owned KMOX-TV St. Louis using modified versions of the French (ANTIOPE) and the British (CEEFAX) teletext system to provide the foundation for the design of a U.S. standard. |
| September 1979 | CBS announces results of first teletext engineering tests. |
| December 1979 | CBS begins feeding teletext on the CBS Television Network to evaluate signal requirements for network transmission. |
| January 1980 | St. Louis tests expand to include broadcast tests over UHF station KDNL-TV St. Louis. During this phase of the test, teletext on KMOX-TV, CBS Owned television station, and KDNL-TV tested. |
| May 1980 | CBS announces results of Phase II testing. |
| June 1980 | Results of broadcast tests over CBS Owned KNXT Los Angeles published in a CBS Engineering Report. |
| July 29, 1980 | CBS files a petition with the Federal Communications Commission for a national broadcast standard for a teletext system based on a modified Antiope system. |

(More)

**CBS EXTRAVISION Brochure. CBS—New York.**

| | |
|---|---|
| November 13, 1980 | The CBS/Broadcast Group, CBS Owned KNXT Los Angeles, public television station KCET, The Caption Center of WGBH in Boston and Telediffusion de France, developer of the teletext system and supplier of the equipment announce plans to participate in a unique program/audience test of teletext in Los Angeles. |
| April 8, 1981 | CBS begins its on-air testing of a teletext news, information and captioning service over KNXT in Los Angeles. The CBS teletext system is called EXTRAVISION.<sup>TM</sup> Fifteen public sites were announced for the on-air test. |
| May 20, 1981 | The French Antiope broadcast teletext system as modified by CBS, the Telidon system developed by the Canadian Department of Communications, and the AT&T standard for videotex coding will be compatible, it is announced at Videotex '81 in Toronto. NBC states it will enter teletext market tests at its station KNBC-TV Los Angeles in Fall 1981. |
| June 22, 1981 | North American Broadcast Teletext Specification (NABTS) standard adopted by CBS for the transmission of the teletext signal.<br><br>This standard is an extension of the French Antiope and Canadian Telidon systems, and is supported by CBS, NBC, and AT&T. |
| March 23, 1982 | CBS begins teletext test in 100 homes in the Los Angeles area. |
| June 25, 1982 | CBS announces plans to launch national teletext service. |
| November 1, 1982 | CBS announces plans to move teletext project from CBS/Broadcast Group to CBS Television Network. Albert H. Crane III named Vice President, Teletext, CBS Television Network, responsible for managing all aspects of the service. |
| March 31, 1983 | The Federal Communications Commission announces that broadcasters may transmit teletext. The FCC ruled not to establish a broadcast standard. |
| April 4, 1983 | CBS launches EXTRAVISION, the broadcast teletext service of the CBS Television Network. EXTRAVISION is broadcast nationally, 24 hours a day, and provides an array of information services as well as a closed-captioning service for deaf and hearing-impaired television viewers. |
| May 15, 1983 | CBS and affiliate WBTV, channel 3 in Charlotte, N.C. (a Jefferson-Pilot Broadcasting Co. station), announce plans to launch the first local station to originate a local EXTRAVISION service. |

(More)

**CBS EXTRAVISION Press Release. CBS—New York.**

| | |
|---|---|
| June 1, 1983 | EXTRAVISION begins its three-month advertising research study. Participants include media, network, and new technology executives from the top 100 advertising agencies. |
| June 28, 1983 | EXTRAVISION demonstrates captioning and program related information pages "live" at a special screening associated with Videotex '83 in New York. |
| September 5, 1983 | EXTRAVISION cumulates positive results of advertising research study. Research indicates agencies like teletext as a means of advertising in overall media mix. |
| January 7, 1984 | Panasonic and Sony announce intentions to begin mass marketing NABTS standard teletext decoders for consumers. Statements were made at the Consumer Electronics Show in Las Vegas. |
| March 5, 1984 | CBS announces the introduction of dual-mode captioning. The alteration to existing policy will expand EXTRAVISION's captioning efforts. |
| April 4, 1984 | CBS affiliate WBTV launches EXTRAVISION service. The Jefferson-Pilot station in Charlotte is the first affiliate in the nation to provide both a network and a local EXTRAVISION service. CBS donates decoders for use of deaf and hearing-impaired community. |
| April 16, 1984 | EXTRAVISION demonstrates service at Videotex '84 in Chicago. |
| April 30, 1984 | WIVB in Buffalo announced intentions to launch a locally originated EXTRAVISION service on July 11, 1984. The statement was made at the NAB Convention in Las Vegas. |
| May 15, 1984 | Bonneville Group stations KSL-TV in Salt Lake and KIRO-TV in Seattle jointly announced intentions to launch locally originated EXTRAVISION services in NABTS. KSL-TV was the first station to broadcast World System teletext and has changed over to the NABTS standard used by CBS, NBC, and AT&T. This commitment has worldwide significance. |
| May 15, 1984 | KCBS in Los Angeles will broadcast a local and network EXTRAVISION service in the third quarter with specially designed Olympic pages. KCBS EXTRAVISION will be demonstrated in 50 public sites beginning July 1, 1984. |

**CBS EXTRAVISION Press Release. CBS—New York.**

**BE THE FIRST TO KNOW**

NBC Teletext's first year of continuous broadcast has been one of growth and improvement in the service. In 1984 our commitment continues with major emphasis placed on consumer promotions, the addition of local teletext services and on-going market research.

The timely information on demand that teletext provides, fills a consumer need as yet unmet by competing media. It helps in daily planning—with weather reports, movie reviews and entertainment guides. And it answers the need to stay informed—with up-to-the-minute news, sports scores and financial reports.

### On Television, On Demand

Broadcast teletext, like broadcast television, is free to the consumer. NBC Teletext will be advertiser-supported—with the potential reach of television, and production costs lower than print. Teletext promises to be a major mass medium in the 1990's. For advertisers, teletext offers an active audience and high impact advertisements selected by the viewer for their information value rather than seen as intrusions. Its color, animation and high quality graphics will make this service a highly competitive advertising vehicle.

In an increasingly active home information marketplace, teletext offers a television-based, mass-market information source programmed to meet the needs and interest of the almost 90 million TV households. It will be differentiated from other services (such as two-way Videotex) by its reach, its mass appeal and its focus on timely programming.

### A strong position in a new industry

NBC Teletext enters this new industry with the strong consumer image and the community presence of NBC and its affiliates. Together we bring nationwide distribution, knowledge of the consumer market for entertainment and information and an effective network for ad sales.

NBC launched network teletext in May 1983. We chose NABTS, with its high-resolution graphics capability, and have created a vivid word and picture format for this new consumer medium, which is exciting to watch and attractive to the home viewer. Our goal for 1984 is to promote, enhance and refine this service. With decoders now commercially available and expected to drop in price later this year, teletext becomes a reality for home viewers. NBC Teletext is actively supporting all manufacturers' efforts to market NABTS decoders.

We are planning a major promotional thrust in conjunction with our affiliate, WDSU, around the 1984 Louisiana World Exposition in New Orleans. NBC Teletext will be featured at WDSU's Telecommunication Center on the fair grounds. With over 10 million

**NBC Teletext Press Release. NBC—New York.**

visitors expected at the fair, a sizeable audience will experience NBC Teletext live. We will also have a presence in Los Angeles, aimed at the many visitors drawn to the 1984 Summer Olympics.

## Local Services Proliferate

Local content offers exciting possibilities for broadcast teletext. Local news, weather, sports, entertainment, shopping specials . . . the opportunities go on. The technology is ready—and equipment is available for local origination. NBC Teletext fully supports affiliates' entry into teletext. We will also be developing local teletext services through our owned stations.

NBC Teletext is growing. Our staff now consists of 20 highly qualified design, editorial, technical and marketing professionals, with additional growth planned to support new local sources.

NBC Teletext has arrived, and we want you to "be the first to know."

For more information on NBC's Teletext service contact:

NBC Teletext
30 Rockefeller Plaza
New York, NY 10020

For information on TV sets and NABTS decoders:

Quasar Company
Franklin Park, Ill. 60131

Panasonic
Secaucus, N.J. 07094

Sony Consumer Products Co.
Park Ridge, N.J. 07656

**NBC Teletext Press Release. NBC—New York.**

Press Department/30 Rockefeller Plaza/New York. N.Y. 10112

## NBC TELETEXT UNVEILS FIRST PAID ADVERTISEMENT AT NAB

* Sony Broadcast Products Company is the first paying sponsor for NBC Teletext, with a specially developed advertisement broadcast on the network service. The three page ad featuring new Sony products is also highlighted on the main index page.

* NBC's Teletext service can be viewed at the Sony booth as well as on decoders demonstrated by Panasonic and other manufacturers. NBC's Hospitality Suite is also featuring the teletext service. Local origination equipment is being demonstrated by VSA at Booth 1001, where NBC will be originating its prototype Las Vegas/NAB magazine.

* The major news about teletext for affiliates is the availability of local equipment for continuous broadcast and local origination.

> -Professional decoders are available from Panasonic and (shortly) from Norpak. These decoders are suitable for monitoring the service in a broadcast environment and will serve as an interface for TV monitors.

> -Data Bridges (which insert network teletext into the local feed) are also now available from EEG and will be available shortly from VSA and Norpak.

> -Local origination equipment providing local magazine storage, page creation capabilities and local insertion is also now available from both VSA and Norpak.

**NBC Teletext Press Release. NBC—New York.**

Press Department/30 Rockefeller Plaza/New York, N.Y. 10112

-Consumer decoders have been announced by Quasar, Panasonic and Sony, with a number of other set manufacturers expected to make announcements this year.

## Also In The News

\* Don't miss the panel on Tuesday, May 1, when Barbara Watson, General Manager, NBC Teletext, will be addressing key teletext issues on a panel headed by industry expert, Gary Arlen. The panel will include Terry Connelly, Taft Broadcasting; and Albert Crane, CBS Extravision.

\* If you're staying for the special session of the NANBA technical committee, Peter Smith, Senior Staff Engineer, NBC Operations and Technical Development, will be discussing "The Status of Teletext in North America" on Thursday, May 3.

Be the first to know: Keep watching NBC Teletext.

**NBC Teletext Press Release. NBC—New York.**

# Broadcasting ⚡ Apr 9

NABTS Teletext Makes Local Debut in Charlotte

NABTS Announcement. Broadcasting Magazine—Washington.

# Birth Announcement

Jefferson-Pilot Broadcasting Company and WBTV in Charlotte are proud to announce the birth of the nation's first NABTS teletext magazine combining national and local data into a single new and innovative information service. This important new industry, now serving Charlotte area viewers, first came to life at 6 pm on April 4, 1984.

We gratefully acknowledge the pioneering and continuing support of our partners, CBS EXTRAVISION,™ and the solid commitment of the system's developer, Videographic Systems of America.

We are pleased to add this new service to a long list of programming firsts dating back to 1922 when WBT Radio signed on as the first radio station in the Carolinas and one of the first in the nation.

**Charlotte:** WBT, WBTV, WBCY, Jefferson-Pilot Teleproductions, Jefferson-Pilot Data Systems, Jefferson-Pilot Retail Services. **Richmond:** WWBT. **Atlanta:** WQXI-FM, WQXI-AM. **Miami:** WGBS, WLYF. **Greensboro:** WBIG. **Denver:** KIMN, KYGO.

WBTV-TV Press Release. WBTV-TV—Charlotte.

WBTV EXTRAVISION Layout for April 4, 1984

| | | | |
|---|---|---|---|
| *00. | Master Index | *53. | Local Services Guide |
| *01. | News Update (Index) | *54. | Road Advisories-1 (Detail) |
| 02. | News-1 | *55. | Road Advisories-2 (Map) |
| 03. | News-2 | *56. | Local Entertainment-1 |
| 04. | News-3 | *57. | Local Entertainment-2 |
| 05. | News-4 | *58. | Local Entertainment-3 |
| 06. | News-5 | *59. | Discovery Place General Informa |
| 07. | News-6 | *60. | Discovery Place Show Schedule |
| 08. | News-7 | *61. | Today on WBTV3 |
| 09. | Full-screen Ad-1 | *62. | Mailbox |
| 10. | Full-screen Ad-2 | 63. | Entertainment Spotlight (Index) |
| *11. | News-1 | 64. | Hollywood Hotline-1 |
| *12. | News-2 | 65. | Hollywood Hotline-2 |
| *13. | News-3 | 66. | Movie Review |
| *14. | News-4 | 67. | Movie Trivia Quiz |
| *15. | News-5 | 68. | Prime Time TV Log-1 |
| *16. | News-6 | 69. | Prime Time TV Log-2 |
| *17. | Weather Watch (Index) | 70. | TV Tonight-1 |
| 18. | U.S. Weather Map | 71. | TV Tonight-2 |
| 19. | U.S. Temps-1 | 72. | Soap Opera Diges:-1 & 2 |
| 20. | U.S. Temps-2 | 73. | Soap Opera Digest-3 & 4 |
| 21. | Full-screen Ad | 74. | Full-screen Ad-1 |
| *22. | Your exclusive WBTV Accu-Weather Forecast | 75. | Full-screen Ad-2 |
| | | 76. | Kaleidescope (Index) |
| *23. | Extended forecast | 77. | Kids-1 |
| *24. | Carolinas' Beaches and Mountains | 78. | Kids-2 |
| *25. | Finance Today (Index) | 79. | Kids-3 |
| 26. | Dow Jones Averages | 80. | Kids-4 |
| 27. | NYSE Ten Most Active | 81. | Kids-5 |
| 28. | Amex Ten Most Active | 82. | Kids-6 |
| 29. | Market Diary | 83. | Kids-7 |
| 30. | Business Briefs-1 | 84. | Full-screen Ad-1 |
| 31. | Business Briefs-2 | 85. | Full-screen Ad-2 |
| 32. | Tax Tips | 86. | Newscaps (Index) |
| 33. | Full-screen Ad-1 | 87. | Newscaps-1 |
| 34. | Full-screen Ad-2 | 88. | Newscaps-2 |
| *35. | Listed Stocks of Local Interest | 89. | Newscaps-3 |
| *36. | O-T-C Stocks of Local Interest | 90. | Caption Guide |
| *37. | Sports Scene (Index) | 91. | Sciencescope (Index) |
| 38. | Sports | 92. | Closeup/Focus-1 |
| 39. | Sports | 93. | Closeup/Focus-2 |
| 40. | Sports | 94. | Science Feature-1 |
| 41. | Sports | 95. | Science Feature-2 |
| 42. | Sports | 96. | Science Feature-3 |
| 43. | Sports | 97. | Update-1 |
| 44. | Special Event | 98. | Update-2 |
| 45. | Special Event | 99. | Full-screen Ad |
| 46. | Sports Quiz | 100. | Masthead |
| 47. | Full-screen Ad-1 | | |
| 48. | Full-screen Ad-2 | | |
| *49. | WBTV3 Sports Spotlight | | |
| *50. | This week's scoreboard | | |
| *51. | WBTV3 Sports Calendar | | |
| *52. | Golf Update | | |

* Frame Created By WBTV

**WBTV-TV EXTRAVISION Layout. WBTV-TV—Charlotte.**

CBS TELEVISION NETWORK
PRESS INFORMATION
51 WEST 52 STREET
NEW YORK, N.Y. 10019

# WIVB-TV, BUFFALO, INAUGURATES LOCAL AND NETWORK EXTRAVISION<sup>TM</sup>

### Second CBS Affiliate Begins Service Providing on Demand
### Information and Captioning, Free Over the Air, in NABTS Standard

WIVB-TV, Buffalo, will become the second CBS Television Network affiliate to inaugurate a local and network EXTRAVISION broadcast teletext service on July 1, it was announced today at the National Association of Broadcasters Convention in Las Vegas by Leslie G. Arries Jr., President of WIVB-TV, and Albert H. Crane III, Vice President, EXTRAVISION Service, CBS Television Network.

The WIVB-TV service will provide 100 "pages" — initially 20 and eventually 50 of them local — of electronically generated information which viewers will receive on demand and free over the air. The service will also provide selected captioned programs for deaf and hearing impaired persons, using an electronic decoder.

This will supplement the "pages" of CBS Television Network service that are broadcast on a 24-hour basis, and updated for 15 hours a day. The CBS EXTRAVISION network service was inaugurated on April 4, 1983, transmitted in the North American Broadcast Teletext Specification (NABTS) with its superior ability to deliver high-resolution graphics. The nation's first local and network EXTRAVISION service began at WBTV, Charlotte, on April 4, 1984.

WIVB-TV will place television sets equipped with decoders in about 10 public locations in the Buffalo area to demonstrate EXTRAVISION. Information and results of events at the Olympics will be constantly updated July 28 through August 13 to demonstrate broadcast teletext's capabilities to the public.

Mr. Arries also announced that his station will utilize Norpak Corporation equipment for origination of local pages. Norpak Corporation is a leading broadcast teletext systems manufacturer, a pioneer in the industry. In addition, Mr. Arries announced that page-creation service will be provided by Macrotel, Inc., of Buffalo.

"We at WIVB-TV are pleased to be the second CBS Television Network affiliate to begin origination of local teletext," Mr. Arries said. "EXTRAVISION is another effort by Channel 4 to continue its leadership role in providing service, information and entertainment to our audience. We look forward to the creative challenge that local EXTRAVISION represents."

"CBS is encouraged," said Mr. Crane, "by our second EXTRAVISION partnership, and we hope to be making more of these announcements throughout what we feel will be the year of awareness for broadcast teletext in our industry — awareness especially of the vital role of the local station, which is the essential backbone of EXTRAVISION. It will also increase our service to the deaf and hearing impaired community."

**WIVB Inaugurates EXTRAVISION Brochure. WIVB-TV—Buffalo.**

WIVB EXTRAVISION Brochure. WIVB-TV—Buffalo.

# WIVB-TV
# EXTRAVISION

## WHAT IS TELETEXT?

Broadcast teletext is a general term for one-way information systems that transmit information over the vertical blanking interval. VBI is the 'black bar' between picture frames seen when the TV screen 'rolls'.

## WHAT IS EXTRAVISION?

The teletext service of the CBS Television Network and WIVB-TV is called "Extravision." It uses the NABTS (North American Broadcast Teletext Specification) broadcast standard, also supported by NBC, AT&T and Group W.

## WHAT DOES EXTRAVISION PROVIDE?

CBS Extravision is a 100-page electronic national magazine, along with a magazine containing 20 pages of local information of particular interest to Western New Yorkers from WIVB-TV, Channel 4. These comprehensive national and local services provide an array of news, weather, sports, finance and feature information constantly accessible 24 hours a day in your home via the TV screen.

## HOW MUCH DOES EXTRAVISION COST?

Extravision is FREE...it is part of the regular Channel 4 signal now coming into your home. You don't need a computer, access lines or cable hookup.

## IF IT IS COMING INTO MY HOME, HOW CAN I SEE EXTRAVISION?

You will need a special TV set and a decoder with a hand-held keypad similar to a TV remote control unit to see Extravision. This decoder unscrambles the signal and gives you Extravision "on demand." The pages appear at the touch of the buttons only when directly accessed by you with the information you want to see when you want to see it.

## ARE THE SETS AND DECODER AVAILABLE NOW, AND WHAT DO THEY COST?

Consumer decoders are currently being manufactured by Panasonic and will soon be available in Buffalo. The NABTS Panasonic decoder will sell initially for approximately $300 and the special TV receiver in the $1000-1200 range.

Quasar, Sony, RCA and Hitachi are also developing models. It is expected that a decoder which attachs to the TV sets now in the home will be available by 1985 and will sell for about $300. Prices are expected to decline as more manufacturers make teletext decoders and consumers become aware of this service. Locations and dealers selling these sets and decoders will be shown on Extravision and Channel 4.

## WHERE CAN I SEE EXTRAVISION?

The Extravision exhibit at the Erie County Fair is the first public display of the service in Buffalo. In the near future, other sites such as malls, supermarkets and retail television and department stores will give Western New Yorkers the opportunity to view it. Watch Channel 4 for these locations.

## WHY HAVEN'T I HEARD OF EXTRAVISION BEFORE?

Extravision is a relatively new technology with CBS launching its service in April of 1983. WIVB-TV in Buffalo is the second of only five cities in the United States who are now or will soon provide local pages. NBC, through WGRZ-TV, is also transmitting 50 pages of national teletext service in Buffalo, but WGRZ is not currently providing any local pages.

## WHY EXTRAVISION?

Besides the television "magazine" information programming that is delivered free to your home 24 hours a day, Extravision is an information and captioning service for the millions of deaf and hearing-impaired viewers. Beginning this fall, CBS will begin dual-mode captioning. There will be a minimum of 3 hours per week of prime-time CBS programs simultaneously closed captioned in the NABTS system and the so-called "line 21" system. This will make it possible for hearing-impaired viewers to receive closed captions through decoders compatible with either format and provide improvements like captioning in different lettering heights and colors to distinguish between actors, multiple speeds and placement of text.

## HOW ARE THE LOCAL EXTRAVISION PAGES CREATED?

Local page creation using high resolution graphics is provided by Macrotel, Inc., of Buffalo. Text information for the pages is gathered and edited by the Channel 4 staff and transmitted on the Channel 4 VBI using Norpak equipment.

### WE'VE GOT THE VISION
### ◉EXTRAVISION™

**WIVB EXTRAVISION Brochure. WIVB-TV—Buffalo.**

EXTRAVISION and the
1985 Auto Show
& Exposition

## The Concept

To join forces with the Niagara Frontier Auto Dealers
Association in introducing the latest broadcast technology
to the people of Western New York.

## The Benefit

EXTRAVISION provides a new medium for dealers and dealer
groups to promote their presence and their products at the
1985 Auto Show & Exposition.

## The Opportunity

WIVB-TV on-air promotional announcements will provide the
NFADA with an excellent opportunity to increase awareness
and interest in the 1985 Auto Show & Exposition.

BUFFALO BROADCASTING CO., INC.   2077 ELMWOOD AVENUE   BUFFALO, NEW YORK 14207   PHONE: 716/874-4410

**WIVB EXTRAVISION Auto Show Brochure. WIVB-TV—Buffalo.**

### The Package

1.  "1985 Auto Show & Exposition" introduction page called up automatically whenever the local index page is requested.

2.  1985 Auto Show pages intermixed with the WIVB Local Information pages, or grouped together in their own section of Auto Show information.

3.  Two styles of pages available:

    A.  Original graphic pages, such as the demonstration pages for the 1985 Chevrolet Spectrum or the Metro Chevy Dealers.  Production cost:  $200 per page.

    B.  "Donut" page containing "1985 Auto Show & Exposition" logo with space for several lines of copy.  Production cost:  $100 per page.

4.  Promotional schedule for EXTRAVISION at the 1985 Auto Show & Expostion.

    Total HH GRPs:       200

    Approx. Value:     $8,000

5.  Two EXTRAVISION kiosks at the Auto Show with a representative present to explain EXTRAVISION and direct viewers to the pages dealing with the Auto Show and its participants.

6.  The 1985 Auto Show pages will be placed on the EXTRAVISION system a week before the Show, and will remain on the system for the week following the Show.

BUFFALO BROADCASTING CO., INC.   2077 ELMWOOD AVENUE   BUFFALO, NEW YORK 14207   PHONE: 716-874-4410

**WIVB EXTRAVISION Auto Show Brochure. WIVB-TV—Buffalo.**

```
          The  Requirements
          ---------------------


   1.  Base fee for EXTRAVISION:                        $1,000
          (Page production charges additional)


   2.  A commitment to WIVB-TV for 40% of the NFADA television
       advertising budget for the 1985 Auto Show & Exposition.
```

BUFFALO BROADCASTING CO., INC.    2077 ELMWOOD AVENUE    BUFFALO, NEW YORK 14207    PHONE: 716/874-4410

**WIVB EXTRAVISION Auto Show Brochure. WIVB-TV—Buffalo.**

# Glossary

| | |
|---|---|
| **Alpha Geometric** | Computer instructions for building a Telidon picture which are based on elementary geometrical forms: line, point, rectangle, arc, polygon. |
| **Alpha Mosaic** | Computer instructions for building a picture from a mosaic of squares. Used by rival Videotex systems. |
| **Baud** | Unit of data transmission speed. One baud is equal to about one bit per second. |
| **Bubble Memory** | An advanced, high speed, high capacity solid state memory. |
| **Byte** | Unit of electromagnetic information or "word". The number of bytes defines a computer's memory capacity, and thus the speed at which it handles data. |
| **Cabletext** | Basic cabletext: Telidon pages converted to video signals by the cable operator and transmitted to all cable subscribers. No terminal is required by the subscriber. Cable teletext: A full cable channel devoted to a teletext cycle of about 20,000 pages which can be selected by subscribers with a terminal. |
| **Decoder** | A Telidon decoder translates electronic information about a transmitted page into a TV signal. |
| **Encoder** | A piece of hardware used to insert Telidon pages in a broadcast signal for Teletext or Cabletext systems. |
| **Fibre Optics** | Technology which uses modulated light signals travelling in a flexible glass fibre as a high speed, high capacity transmission medium. |
| **Hardware** | The physical units connected with a Telidon system—terminal, information provider, etc. |
| **High Resolution Display** | A special screen which provides pictures of much higher quality than a normal TV set. |
| **Page Creation Terminal** | Hardware used to create Telidon pages. |
| **Interface** | Electronic connection between two different pieces of hardware. |
| **Modem** | Hardware used to interface a transmission medium to a Telidon terminal. A telephone modem converts digital data into analog sound that travels along a telephone line. From MODulator - DEModulator. |

# Glossary

| | |
|---|---|
| **Microprocessor** | A miniaturized, low-cost computer. |
| **Non Volatile Memory** | A memory system whose contents are preserved when the device is switched off. |
| **Pixel** | Smallest controllable element that can be illuminated on a display screen. |
| **RAM** | Random access memory. Information may be obtained without having to read sequentially through a list. |
| **Software** | The programs and electronic instructions of a system. |
| **Stand Alone** | A Telidon terminal need not be attached to a transmission medium. It serves as a portable or display unit but has no interactive exchange with a data bank. |
| **Telidon** | Trademark (certification mark) of Canada's Videotex system. |
| **Telidon Page** | A complete unit of data which is defined by a page number. It generally consists of a complete frame but may also comprise a sequence of frames. |
| **Teletext** | Interactive television system that receives its pages by television broadcast. Users cannot interact directly with a data bank. |
| **Videodisc** | Disc recording of sound and images that is read by laser beam or magnetic head. |
| **Videotex** | Two way interactive television system. |

**Glossary of Teletext Terms. Canadian Department of External Affairs—Ottawa.**

# Notes

## INTRODUCTION

1. Joseph Roizen, "Teletext—A Service That's Coming of Age," *Educational and Industrial TV* (September 1982): pp. 39–40. See also Gary Arlen, "Getting Ready for Videotex, Teletext and the World Beyond," *Cable Vision* (1 June 1981): pp. 233ff. Morris Edwards, "Videotex/Teletext Poised for Major Growth in the United States," *Communications News* (August 1982): pp. 88–92.

## CHAPTER 1. TELETEXT: A NEW IDEA FOR NEWS DISSEMINATION

1. "Who Will Dominate the Home of the Future?" *Cable Communications* (August 1982): pp. 18–19. See also Victor Livingston, "Teletext: It's Coming Soon to Home Screens," *Adweek East* (28 April 1982): Sec. 2, p. 48. Martin Mayer, "Coming Fast: Services through the TV Set," *Fortune* (14 November 1983): p. 50.
2. "Components of a Broadcast Videotex System—Teletext," *Videotex/Teletext: Principles & Practices* (New York: McGraw-Hill, 1985), pp. 135–91.
3. "Teletext and the Broadcast Revolution," *North American Teletext: Reaching New Markets* (Ottawa: Department of External Affairs, 1981), pp. 1–6. See also T. J. Logue, "Teletext: Towards an Information Utility," *Journal of Communications* 29, no. 4 (1979): pp. 58–65.
4. Merrill Weiss and Ronald Lorentzen, "How Teletext Can Deliver More Services and Profits," *Broadcast Communications* (August 1982): pp. 54–60.
5. "Teletext and Prestel—User Reactions—Report One—Teletext Users," (London: Communications Studies and Planning Ltd., 1981), pp. 1–17.
6. Ibid., pp. 4–9.
7. Udayan Gupta, "Teletext Attracts New Players as It Combats an Identity Crisis," *Electronic Media* (11 February, 1983): p. 33.
8. "What's Going On in Teletext and Videotex," *Educational and Industrial TV* (September 1982): pp. 42–44.

## CHAPTER 2. TELETEXT DEVELOPMENTS IN EUROPE

1. Richard Hooper, "The British Viewdata and Teletext Standard," *Videotex '82 Proceedings, New York* (London: Online Conferences, Ltd., 28–30 June 1982), p. 414.
2. Ibid., pp. 414–15.
3. Ibid., pp. 418–20.

4. Ibid., p. 420.

5. "Teletext and Prestel—User Reactions—Report One," pp. 8–9.

6. Ibid., pp. 14–17.

7. Colin McIntyre, "Broadcast Teletext—Who Says It Isn't Interactive," *Videotex '82 Proceedings, New York* (London: Online Conferences, Ltd., 28–30 June 1982), pp. 1–12.

8. Ibid., pp. 2–12.

9. Dick Lutz, "Teletext and Videotex Converging," *Videotex '82, New York* 28 June 1982, p. 14.

10. McIntyre, "Broadcast Teletext," p. 9.

11. Ibid., pp. 37–38.

12. Richard Hooper, "The UK Scene—Teletext and Videotex," *Videotex '81 Proceedings, Toronto* (London: Online Conferences, Ltd., 20–22 May 1981), pp. 131–35.

13. Humphrey Metzgen, "Making Money from Teletext," *Videotex '82 Proceedings, New York* (London: Online Conferences, Ltd., 28–30 June 1982), pp. 37–43.

14. Brian G. Champness, "Social Uses of Videotex and Teletext in the U.K.," *Videotex '81 Proceedings, Toronto* (London: Online Conferences, Ltd., 20–22 May 1981), p. 334.

15. McIntyre, "Broadcast Teletext," pp. 10–11.

16. Metzgen, "Making Money," p. 38.

17. McIntyre, "Broadcast Teletext," p. 33.

18. "Business Teletext," *Videotex Viewpoint* (April/May 1985): pp. 33–34.

19. Ibid.

20. Metzgen, "Making Money," pp. 38–40. See also "British Viewers Prefer Teletext to Evening News," *Multichannel News* (9 April 1982): p. 85.

21. McIntyre, "Broadcast Teletext," p. 34.

22. Pierre Gaujard, "State of the Art—France," *Inside Videotex Proceedings* (Toronto: Infomart, 13–14 March, 1980), pp. 26–42.

23. Ibid. See also Jan Gecasi, *The Architecture of Videotex Systems* (Englewood Cliffs, NJ: Prentice-Hall, 1983), pp. 65–69.

24. Paul McFarland, "State of the Art—Britain," *Inside Videotex Proceedings* (Toronto: Infomart, 13–14 March 1980), pp. 20–24.

25. T. Ohlin, "Videotex and Teletext in Sweden—A Nation Decides," *Viewdata '81 Proceedings* (London: Online Conferences, Ltd., October 1981), pp. 215–30.

26. Ibid., p. 225.

27. P. Ferenczy, "The Development of Teletext and Viewdata in Hungary," *Viewdata '81 Proceedings* (London: Online Conferences, Ltd., October 1981), p. 185.

28. Kenneth Edwards, "Broadcast Teletext: The Next Mass Medium?" *The Futurist* (October 1982): pp. 21–24.

29. Tetsuro Tomita, "Japan: The Search for a Personal Information Medium," *Intermedia* 7, no. 3 (1979): pp. 10–18.

## CHAPTER 3: DEVELOPMENTS IN NORTH AMERICA

1. J. R. Storey, A. Vincent and R. Fitzgerald, *A Description of the Broadcast Telidon System* (Ottawa: Federal Department of Communications, 1980), pp. 1–6. See also Nicole F. LeDuc, "Teletext and Videotext in North America: The Canadian Perspective," *Telecommunications Policy* 4, no. 1 (1980): pp. 9–16.

2. J. R. Storey, H. G. Brown, C. D. O'Brien and W. Sawchuk. "An Overview of Broadcast Teletext Systems for NTSC Television Standards" (Ottawa: Federal Department of Communications, 1980), pp. 1–21.

3. *Telidon Reports,* No. 11 (Ottawa: Department of Communications, March 1984), pp. 9–11.

4. *Telidon Reports,* No. 12 (Ottawa: Department of Communications, October 1984), p. 8.

5. *Telidon Reports,* No. 2 (Ottawa: Department of Communications, June 1980), p. 1.

6. *Telidon Reports,* No. 10 (Ottawa: Department of Communications, March 1983), pp. 5–6. See also "Telidon in TV Ontario," *SMPTE Journal,* May 1982, p. 469.

7. *Telidon Reports,* No. 7 (Ottawa: Department of Communications, August 1981), p. 15.

8. *Telidon Reports,* No. 10, pp. 4–5.

9. "Telidon Chosen for First U.S. Consumer Trial of Teletext," news release (Ottawa: Department of Communications, 6 June 1980), pp. 1–2.

10. *Telidon Reports,* No. 9 (Ottawa: Department of Communications, June 1982), p. 11.

11. Jonathan Cheyreau, "Radio Teletext Eliminates Telephone Costs," *The Globe and Mail,* 6 October 1983, p. 22.

12. David Vermilyea, "Where Will Teletext Be in 2000 A.D.?" *Educational and Industrial TV* (July 1980): p. 23.

13. "Teletext Experimentation in the U.S.," *Electronic Publishing in the Home of the 1980s's* (New York: Donaldson, Lufkin & Jenrette, 1981), pp. 1–4.

14. Ibid., pp. 3–4.

15. "Teletext User Survey—Ceefax and Oracle User Reactions" (Washington: Videotex Industry Association, 1982), pp. 1–23.

16. Ibid., p. 23.

17. Philip Keirstead, "Teletext Is in Your Future," *Broadcast Communications* (August 1982): p. 30.

18. Ibid., p. 31.

19. William Fretts, "An Introduction to the NABTS Teletext Terminal," *Videotext World* (December 1985): pp. 37–41.

20. "A Panel Study of Family Teletext Viewing Habits and Preferences" (Washington: National Association of Broadcasters, 1983), pp. 1–5.

21. "Viewer Usage Study." *Electronic Publisher* (16 January 1984): p. 3.

22. "Louisville Teletext Test Begins with 150 Homes," *Multichannel News* (19 April 1982): p. 39.

23. R. C. Morse, "The Home Information Explosion—Part II," *Marketing Communications* (March 1982): pp. 24–25. See also "Keycom Completes Test of Night-Owl," *Cable Vision* (20 September 1982): p. 24.

24. Brad Johnson, "Knight-Ridder Plans Teletext for Detroit," *Electronic Media* (26 August 1982): p. 36.

25. Morse, "The Home Information," p. 25.

26. Ibid., p. 25.

27. Ibid., pp. 25–27.

28. Ibid., p. 28.

29. Ibid.

30. Ibid., p. 71.

31. Ibid.

32. "VSA Launches American Campaign," *Broadcasting* (24 January 1983): p. 16. See also "French Consortium to Install CBS, NBC Teletext Systems," *Communications Daily* (24 January 1983): p. 30. David Percelay, "Strategic Planning for a Major Market Trial of Broadcast Teletext," *Videotex '81 Proceedings, Toronto* (London: Online Conferences, Ltd., 22 May 1981), pp. 71–73.

33. "CBS, NBC Set to Launch National Teletext Service," *Multichannel News*, 5 July 1982, p. 12.

34. "Two Nets Taking Leap into Teletext," *Electronic Media* (8 July 1982): p. 18.

35. "NBC Pact Sets Stage to Launch a National Teletext Magazine," *Electronic Engineering Times* (February 1982): p. 56.

36. Les Luchter, "Teletext Study Rewarding to CBS, NBC," *Broadcast News* (7 March 1983): p. 40. See also "CBS Going Live With Teletext in L.A. Test," *Broadcasting* (17 November 1980): p. 21.

37. "Results of Los Angeles Teletext Research Project," press release (New York: CBS Television Network, September 1982), pp. 1–3.

38. "NABTS—What It Can Mean to Teletext/Videotex," *Educational and Industrial TV* (September 1982): p. 41.

39. Ibid., p. 44.

## Chapter 4. Teletext is Born in the U.S.

1. John Madden, "Telidon Evolution and Future Prospects," (paper presented at the American Association for the Advancement of Science Conference, 7 January 1981), pp. 1–4.

2. W. S. Ciciora, "Virtext and Virdata, A Present U.S. Teletext Application," *Videotex '81 Proceedings, Toronto* (London: Online Conferences, Ltd., May 1981), pp. 77–84.

3. Diane Mermigas, "Keyfax Gets Ready for National Teletext Launch," *Electronic Media* (August 1982): pp. 19–20. See also "Keyfax First Nationally but Only the Beginning," *Cable Age* (31 January 1983): p. 31. "Keyfax Launches 24 Hours Magazine," *Cable Vision* (22 November 1982): p. 70. "Keyfax National Teletext Launches," *Cable Vision* (6 December 1982): p. 10. "Keyfax Teletext Service to Launch November 15," *Cable Age* (6 September 1982): p. 106. "Keyfax Set for Launch in November," *Multichannel News* (23 August 1982): p. 4.

4. Diane Mermigas, "Keyfax Teletext Magazine Set for Debut in November," *Electronic Media* (22 November 1982): p. 4. See also "SSS Field Plan Teletext Service," *Multichannel News* (15 March 1982): p. 1. "Teletext Venture Discussed by SSS, Field Electronics," *Cable Age* (22 March 1982): p. 31.

5. Ibid., p. 5. See also Richard Zacks, "Teletext Service Means Business," *Multichannel News* (20 November 1982): p. 9.

6. Mermigas, "Keyfax Gets Ready," pp. 4 & 16.

7. Ernest Holsendolph, "Teletext Authorized by F.C.C." *New York Times*, April 1, 1983, pp. D10–11.

8. Ron Merrell, "Teletext: New Rules, New Hurdles," *Broadcast Communications* (August 1983): p. 8.

9. Sammy R. Danna, "Keyfax Takes to Videotex," *Educational and Industrial TV* (November 1985): pp. 51–53. See also Anne Dukes, "SSS Ends Involvement in Keyfax Teletext Service," *Multichannel News* (19 November 1984): p. 12. "Teletext Gets Boost With Taft-SSS Venture," *Broadcasting* (4 February 1983): p. 85.

10. Terry Connelly, "Teletext Enhances WKRC's Local News Image," *Television Broadcast Communications* (October 1983): pp. 52–58.

11. "Zenith to Have Advanced Teletext Decoder in 1984," *Multichannel News* (September 9, 1983): p. 4. See also "Taft Zenith Announce Teletext Hardware Deal," *Multichannel News* (January 10, 1983): p. 1.

12. William L. Thomas, *Full Field Tiered Addressable Teletext* (Glenview, Ill.: Zenith Radio Corporation, 1982), pp. 1–9.

13. Ibid., p. 8.

14. David Klein, "Taft's Electra Stars with Breaking News," *Electronic Media* (October 13, 1983): p. 16.

15. "VSA to Promote NABTS: Announces NBC Contract," *Broadcast Management Engineering* (March 1983): pp. 35–36.

16. "NBC Teletext—A History," press release, 1983.

17. "NBC Launches Its Network Service," press release, 1983.

18. "NBC's Barbara Watson," press release, 1983.

19. "NBC Teletext Currently Offering," press release, 1983.

20. Diane Mermigas, "Decoder Production Slows Teletext Service," *Electronic Media* (14 July 1983): pp. 1–2.

21. "CBS to Launch Teletext in April," *Cable Vision* (14 February 1983): p. 36.

22. "NABTS Terminals to Appear, More on the Way," *Videoprint* (9 January 1984): p. 1.

23. Andrew Pollack, "Teletext is Ready for Debut," *New York Times*, 18 February 1983, p. D6.

24. Sidney Shaw, "Teletext Service Will Be Available by Next Year," *Videoprint* (3 May 1984): pp. 11–12.

25. Kenneth Clark, "CBS Broadcast of Text is 1st Shot in TV Revolution," *Charlotte News*, 2 June 1983, p. B1.

26. "CBS Starts Its Teletext Service," *New York Times*, 5 April 1983, p. D5.

27. "Teletext Launched on CBS Affiliate," *Broadcasting* (9 April 1984): p. 37.

28. Jan Margaret Parr, "Can Teletext Hit Home?" *SAT* (April 1984): pp 8–10.

29. Jack Loftus, "CBS, NBC Roll Out Teletext," *Variety* (25 January 1984): pp. 16–17.

30. "Teletext Launched on CBS Affiliate," *Broadcasting* (9 April 1984): p. 37. See also "Nation's First Local and Network Extravision Teletext Service Inaugurated April 4 by WBTV, Charlotte, and CBS Television Network," *CBS Press Release*, 4 April 1984, pp. 1–2.

31. "CBS Affiliate WBTV Charlotte Begins Local NABTS Teletext," *Communications Daily* (5 April 1984): p. 1.

32. Loftus, "CBS, NBC," p. 17. See also "Charlotte to Get NABTS Teletext," *Videoprint* (9 January 1984): pp. 13–15.

33. "Teletext Launched," *Broadcasting*, p. 33. See also Mark Wolf, "CBS Station Debuts Local Teletext Service," *Electronic Media* (12 April 1984): pp. 24–25.

34. Arthur Unger, "Big Broadcasters Join Up to Push Their Version of Teletext TV News Service," *Christian Science Monitor*, 5 April 1984, pp. 20–21.

35. "WIVB Starts Extravision," *Broadcasting* (1 May 1984): p. 4.

36. Diane Mermigas, "Taft Starts Cincinnati Teletext Service," *Electronic Media* (30 June 1983): p. 15.

37. Loftus, "CBS, NBC," p. 17.

38. "CBS to Expand Existing Captioning Efforts for Hearing Impaired through Its Extravision Teletext Service," *CBS Press Release*, 5 March 1984.

39. "Affiliates Exhorted to Support Extravision," *Broadcasting* (21 May 1984): pp. 37–38.

40. "CBS Pushes Extravision," *Videoprint* (8 June 1984): p. 40.
41. "Bonneville International Corporation to Provide CBS Extravision Services," CBS press release, 21 May 1984. See also Louis Chunovic, "Bonneville Chooses NABTS," *Broadcast Week* (21 May 1984): pp. 25–26.
42. "Bonneville Goes with Extravision for Salt Lake, Seattle TV Stations," *International Videotex Teletex News* (June 1984): pp. 5–6.
43. Eileen Norris, "Teletext Goes Local," *Electronic Media* (April 1986): pp. 18–22.
44. Nancy Tracewell, "There's Plenty of News in Those Black TV Lines," *Business First of Buffalo* (19 November 1984): p. 10.
45. "Crane Gives Extravision Update," *Broadcasting* (30 May 1983): p. 29.
46. Kenneth Clark, "Birth of Extravision: CBS Broadcasts a New Electronic 'Wizard' Called Teletext," *Chicago Tribune,* 17 May 1983, Sec. ff.
47. "Random Thoughts," *Broadcast Week* (14 May 1984): p. 56.
48. Merrill Weiss and Ronald Lorentzen, "How Teletext Can Deliver More Service and Profits," *Broadcast Communications* (August 1982): p. 10. See also Roizen, "Teletext—A Service That's Coming of Age," p. 40.
49. Martin Lane, "New Life in the Airwaves: Broadcasting Business Data," *Videotex World* (June 1985): pp. 9–13.
50. "Role Seen for 'Teletext' in Home Banking," *EFT Report* (January 1983): p. 3.
51. "Merrill Lynch, PBS to Devise 3-City Teletext Demo Project," *Washington Post,* 10 April 1983, pp. 6–7. See also "Explore Teletext's Use," *Wall Street Journal,* 5 April 1983, p. 43.
52. "Satellite-Delivered Text Service Signs 4 Carriers," *Multichannel News* (18 June 1984): p. 18.

## CHAPTER 5. PRINT MEDIA BECOMES A MAJOR PARTICIPANT

1. Johnson, "Knight-Ridder Plans Teletext for Detroit," p. 36.
2. "Teletext to Threaten Papers, Newscasts, IRD Study Predicts," *Multichannel News* (3 September 1984): p. 59. See also "Teletext: TV Gets Married to the Printed Word," *Broadcasting* (20 August 1979): pp. 30–35.
3. *Telidon Reports,* No. 5 (Ottawa: Department of Communications, May 1981), p. 1.
4. "Consumer Communications Center," *Time Brochure,* 1981, pp. 1–2.
5. John Lopinto, "Time Incorporated National Teletext Service," *Videotex '81 Proceedings, Toronto* (London: Online Conferences, Ltd., 20–22 May 1981), pp. 109–12.
6. "Time Plans Launch of Teletext Trial," *Multichannel News* (11 October 1982): pp. 1 & 12. See also "Time Begins Teletext Test," *Cable Vision* (8 November 1982): p. 69. "Time Teletext Sets Out to Create a New Medium," *Cable Age* (17 January 1983): p. 26.
7. Jeri Baker, "Out of the Time Incubator," *Cablevision Plus* (22 March 1982): pp. 17–18.
8. Ibid., p. 17.
9. Ibid.
10. Ibid., p. 19.
11. Ibid.
12. Ibid., p. 17.

13. Ibid., p. 18.
14. Ibid., p. 19.
15. Ibid., p. 20.
16. Larry T. Pfister, "Teletext . . . Its Time Has Come," *National Cable Television Association Conference, Las Vegas,* 2 May 1982, p. 1.
17. Ibid., p. 4.
18. "Time, Inc. Adopts North American Broadcast Teletext Standard for Its National Service," *Time Incorporated News Release,* 20 November 1981, pp. 1–2.
19. *Telidon Reports,* No. 5, p. 2.
20. Pfister, "Teletext . . . ," p. 7.
21. Ibid.
22. Deborah Wise, "Time Teletext to Use Panasonic Terminals," *Infoworld* (November 1982): pp. 22–24. See also Susan Spillman, "Time, Matsushita Link on Teletext Terminals," *Electronic Media* (16 December 1982): p. 20. Laura Landro, "Time Inc.'s Teletext Will Use Terminals Made by Matsushita," *Wall Street Journal,* 14 December, 1982, p. 10.

## Chapter 6. Viewer Impressions

1. Data collected from Business Telecommunications Workshop conducted in July 1985 at Buffalo State College. The participants included nineteen graduate students majoring in business.
2. Ibid.
3. Melanie Herberger Collins, "Teletext and Videotex: New Services through the Television Set," college term paper, 15 November 1984, p. 7.

## Chapter 7: Starting a Local Teletext Service

1. Alan Pergament, "Channel 4 Starts Electronic Newspaper," *Buffalo News,* 2 August 1984, p. B-4.
2. Letter from Macrotel to WIVB—11 February 1983.
3. Letter from Graziplene to Norpak—8 January 1984.
4. Letter from Graziplene to Arries, Crane, and Norton—28 February 1984.
5. Letter from Macrotel to WIVB—18 September 1984.
6. "Tampa Hosts A Television Startup," *Videoprint* (8 October 1985): pp. 4–5.
7. Mary Fialkiewicz, "TV Turns the Page," *Niagara Gazette,* 9 March 1986, pp. 26–27.
8. Albert R. Crane III, "Extranet Inc.—A New Television Data Broadcasting Corporation Formed," *Press Release,* 1 August 1986.
9. Suzanne Aberbach, "Those Miserable Lines Marring TV Bring Smiles to This Businessman," *Bronxville Review Press Reporter,* 20 November 1986, Section A1.
10. Ibid.
11. *Extranet Brochure,* 1986.

## Chapter 8. The Failure of Teletext

1. "Status of the Videotex Industry," *Videoprint* (10 January 1983): p. 1+.
2. Michael R. Raggett, "Teletext's Uncertain Future," *Computer Graphics World* (February 1985): pp. 73–74.

3. Udayan Gupta, "Huttonline Offers Ambitious Electronic Investment Service," *Electronic Media* (12 January 1984): p. 19.

4. "Agricultural Teletext Demonstration—Infotext," *University of Wisconsin Extension,* March 1986, pp. 1–17.

5. Ibid., p. 4.

6. Ibid., pp. 12–14.

7. Ibid., p. 6.

8. Ibid., pp. 16–21.

9. "Broadcast Teletext: Implications of a Scarce Resource," *Videotex Viewpoint* (April/May 1985): chapter 3.

10. R. R. Bruce, "US Legal and Regulatory Issues Relating to Teletext and *Videotex*—A Layman's Guide to a Lawyer's 'No-Man's Land'," *Viewdata '81 Proceedings, Toronto* (London: Online Conferences, Ltd., October 1981), pp. 367–83.

11. "Teletext Finding Wide Acceptance," *Broadcasting* (February 1987): p. 21.

12. Nicholas Timasheff, *Sociological Theory: Its Nature and Growth,* 3d ed. (New York: Random House, 1967), pp. 207–8.

# Bibliography

"A Panel Study of Family Teletext Viewing Habits and Preferences." National Association of Broadcasters, 1983, pp. 1–5.

Aberbach, Suzanne. "Those Miserable Lines Marring TV Bring Smiles to This Businessman." *Review Press Reporter, Bronxville, N.Y.*, 20 November 1986, Section A.

"Agricultural Teletext Demonstration." *University of Wisconsin Extention*, March 1986, pp. 1–17.

Arlen, Gary. "Getting Ready for Videotex, Teletext and the World Beyond." *Cable Vision* (1 June 1981): p. 233ff.

Baker, Jeri. "Out of the Time Incubator." *Cablevision Plus* (22 March 1982): pp. 17–20.

"Bonneville Goes with Extravision for Salt Lake, Seattle TV Stations." *International Videotex Teletext News* (June 1984): pp. 5–6.

"Bonneville International Corporation to Provide CBS Extravision Services." *Press Release, CBS Broadcast Group*, 21 May 1984.

*Broadcasting.* 20 August 1979: "Teletext: TV Gets Married to the Printed Word," pp. 30–35.

———. 17 November 1980: "CBS Going Live with Teletext in L.A. Test," p. 21.

———. 24 January 1983: "VSA Launches American Campaign," p. 16.

———. 4 February 1983: "Teletext Gets Boost With Taft-SSS Venture," p. 85.

———. 30 May 1983: "Crane Gives Extravision Update," p. 29.

———. 9 April 1984: "Teletext Launched on CBS Affiliate," p. 37.

———. 1 May 1984: "WIVB Starts Extravision," p. 4.

———. 7 May 1984: "WIVB-TV Plans Teletext Launch."

———. 21 May 1984: "Affiliates Exhorted to Support Extravision," pp. 37–38.

———. 26 May 1984: "High Sights, Low Visibility," pp. 36–46.

———. 18 February 1987: "Teletext Finding Wide Acceptance," p. 21.

Bruce, R. R. "US Legal and Regulatory Issues Relating to Teletext and Videotex—A Layman's Guide to a Lawyer's 'No-Man's Land'." *Viewdata '81 Proceedings*. Online Conferences, London, UK, October 1981, pp. 367–83.

*Cable Age.* 22 March 1982: "Teletext Venture Discussed by SSS, Field Electronics," p. 31.

———. 6 September 1982: "Keyfax Teletext Service to Launch November 15," p. 106.

———. 17 January 1983: "Time Teletext Sets Out to Create a New Medium," p. 26.

———. 31 January 1983: "Keyfax First Nationally but Only the Beginning," p. 31.

*Cable Vision.* 20 September 1982: "Keycom Completes Test of Night-Owl," p. 24.

———. 8 November 1982: "Time Begins Teletext Test," p. 69.

———. 22 November 1982: "Keyfax Launches 24 Hours Magazine," p. 70.

———. 6 December 1982: "Keyfax National Teletext Launches," p. 10.

———. 14 February 1983: "CBS to Launch Teletext in April," p. 36.

"CBS Starts Its Teletext Service." *New York Times,* 5 April 1983, p. D5.

"CBS to Expand Existing Captioning Efforts for Hearing Impaired through Its Extravision Teletext Service." *Press Release, CBS Broadcast Group,* 5 March 1984.

Champness, Brian G. "Social Uses of Videotex and Teletext in the U.K." *Videotex '81 Proceedings, Toronto.* Online Conferences, Ltd., London, UK, 20–22 May 1981, pp. 331–39.

Cheyreau, Jonathan. "Radio Teletext Eliminates Telephone Costs." *Globe and Mail,* 6 October 1983, p. 22.

Chunovic, Louis. "Bonneville Chooses NABTS." *Broadcast Week* (21 May 1984): pp. 25–26.

Ciciora, W. S. "Virtext and Virdata, A Present U.S. Teletext Application." *Videotex '81 Proceedings, Toronto.* Online Conferences, Ltd., London, UK, 20–22 May 1981, pp. 77–84.

Clark, Kenneth R. "Birth of Extravision: CBS Broadcasts a New Electronic 'Wizard' Called Teletext." *Chicago Tribune,* 17 May 1983, Sec. ff.

Clark, Kenneth. "CBS Broadcast of Text is 1st Shot in TV Revolution." *Charlotte News,* 2 June 1983, p. B1.

*Communications Daily.* 24 January 1983: "French Consortium to Install CBS, NBC Teletext Systems," p. 30.

———. 5 April 1984: "CBS Affiliate WBTV Charlotte Begins Local NABTS Teletext," p. 1.

"Components of a Broadcast Videotex System—Teletext." *Videotex/Teletext: Principles & Practices* (New York: McGraw-Hill, 1985), pp. 135–91.

Connelly, Terry. "Teletext Enhances WKRC's Local News Image." *Television Broadcast Communications* (October 1983): pp. 52–58.

"Consumer Communications Center." *Time Brochure,* 1981, pp. 1–2.

Crane, Albert R. III, "Extranet, Inc.—A New Television Data Broadcasting Corporation Formed." *Press Release,* 1 August 1986.

Danna, Sammy R. "Keyfax Takes to Videotex." *Educational and Industrial TV* (November 1985): pp. 51–53.

Dukes, Anne. "SSS Ends Involvement in Keyfax Teletext Service." *Multichannel News* (19 November 1984): p. 12.

*Educational and Industrial TV.* September 1982: "What's Going on in Teletext and Videotex," pp. 42–44.

———. September 1982: "NABTS—What It Can Mean to Teletext/Videotex," p. 41.

Edwards, Kenneth. "Broadcast Teletext: The Next Mass Medium?" *The Futurist* (October 1982): pp. 21–24.

Edwards, Morris. "Videotex/Teletext Poised for Major Growth in the United States." *Communications News* (August 1982): pp. 88–92.

"Explore Teletext's Use." *Wall Street Journal,* 5 April 1983, p. 43.

*Extranet Brochure,* 1986.

Ferenczy, P. "The Development of Teletext and Viewdata in Hungary." *Viewdata '81 Proceedings.* Online Conferences, Ltd., London, UK, October 1981, pp. 179–85.

Fialkiewicz, Mary. "TV Turns the Page." *Niagara Gazette,* 19 March 1986, pp. 26–27.

Fretts, William. "An Introduction to the NABTS Teletext Terminal." *Videotex World* (December 1985): pp. 37–41.

Gaujard, Pierre. "State of the Art—France." *Inside Videotex Proceedings.* Toronto: Infomart, 13–14 March 1980, pp. 26–42.

Gecasi, Jan. *The Architecture of Videotex Systems.* Englewood Cliffs, NJ: Prentice-Hall, 1983.

Gupta, Udayan. "Huttonline Offers Ambitious Electronic Investment Service." *Electronic Media* (12 January 1984): p. 19.

———. "Teletext Attracts New Players as It Combats an Identity Crisis." *Electronic Media* (11 February 1983): p. 33.

Herberger Collins, Melanie. "Teletex and Videotex: New Services through the Television Set." College report, 15 November 1984, p. 7.

Holsendolph, Ernest. "Teletext Authorized by F.C.C." *New York Times,* 1 April 1983, pp. D10–11.

Hooper, Richard. "The UK Scene—Teletext and Videotex." *Videotex '81 Proceedings, Toronto.* Online Conferences, Ltd., London, UK, 20–22 May 1981, pp. 131–35.

———. "The British Viewdata and Teletext Standard." *Videotex '82 Proceedings.* Online Conferences, Ltd., London, UK, 28–30 June 1982, pp. 413–21.

Jackson, Charles L., Harry M. Shooshan III, and Jane L. Wilson, "Broadcast Teletext: Implications of a Scarce Resource." *Modern Media Institute,* St. Petersburg, Fla., 1981, pp. 15–23.

Johnson, Brad. "Knight-Ridder Plans Teletext for Detroit." *Electronic Media* (26 August 1982): p. 36.

Keirstead, Philip. "Teletext Is in Your Future." *Broadcast Communications* (August 1982): pp. 30–31.

Klein, David. "Taft's Electra Stars with Breaking News." *Electronic Media* (13 October 1983): p. 16.

Landro, Laura. "Time Inc.'s Teletext Will Use Terminals Made by Matsushita." *Wall Street Journal,* 14 December 1982, p. 10.

Lane, Martin. "New Life in the Airwaves: Broadcasting Business Data." *Videotex World* (June 1985): pp. 9–13.

LeDuc, Nicole F. "Teletext and Videotex in North America: The Canadian Perspective." *Telecommunications Policy* 4, no. 1 (1980): pp. 9–16.

*Letter:* Macrotel to WIVB, 11 February 1983.

———. Graziplene to Norpak, 8 January 1984.

———. Graziplene to Aries, Crane, and Norton, 28 February, 1984.

———. Macrotel to WIBV, 18 September 1984.

Livingston, Victor. "Teletext: It's Coming Soon to Home Screens." *Adweek East* (28 April 1982): Sec. 2, p. 48.

Loftus, Jack. "CBS, NBC Roll Out Teletext." *Variety* (25 January 1984): pp. 16–17.

Logue, T. J. "Teletext: Towards an Information Utility." *Journal of Communications* 29, no. 4 (1979): pp. 58–65.

Lopinto, John. "Time Incorporated National Teletext Service." *Videotex '81 Proceedings, Toronto.* Online Conferences, Ltd., London, UK, 20–22 May 1981, pp. 109–12.

Luchter, Les. "Teletext Study Rewarding to CBS, NBC." *Broadcast News* (7 March 1983): p. 40.

Lutz, Dick. "Teletext and Videotex Converging." *Videotex '82, New York.* 28 June 1982, p. 14.

Madden, John. "Telidon Evolution and Future Prospects." Paper presented at the American Association for the Advancement of Science Conference, 7 January 1981, pp. 1–4.

Mayer, Martin. "Coming Fast: Services Through the TV Set." *Fortune* (14 November 1983): p. 50.

McFarland, Paul. "State of the Art—Britain." *Inside Videotex Proceedings.* Toronto: Infomart, 13–14 March 1980, pp. 20–24.

McIntyre, Colin. "Broadcast Teletext—Who Says It Isn't Interactive." *Videotex '82 Proceedings,* Online Conferences, Ltd., London, UK, 28–30 June 1982, pp. 1–12, 33 & 37–38.

Mermigas, Diane. "Keyfax Gets Ready for National Teletext Launch." *Electronic Media* (August 1982): pp. 4, 16 & 19–20.

———. 22 November 1982: "Keyfax Teletext Magazine Set for Debut in November." *Electronic Media,* pp. 4–5.

———. 30 June 1983: "Taft Starts Cincinnati Teletext Service." *Electronic Media,* p. 15.

———. 14 July 1983: "Decoder Production Slows Teletext Service." *Electronic Media,* pp. 1–2.

Merrell, Ron. "Teletext: New Rules, New Hurdles." *Broadcast Communciations* (August 1983): p. 8.

"Merrill Lynch, PBS to Devise 3-City Teletext Demo Project." *Washington Post,* 10 April 1983, pp. 6–7.

Metzgen, Humphrey. "Making Money From Teletext." *Videotex '82 Proceedings, New York City.* Online Conferences, Ltd., London, UK, 28–30 June 1982, pp. 37–43.

Morse, R. C. "The Home Information Explosion—Part II." *Marketing Communications* (March 1982): pp. 24–29 & 71.

*Multichannel News.* 15 March 1982: "SSS Field Plan Teletext Service," p. 1.

———, 8 April 1982: "British Viewers Prefer Teletext to Evening news," p. 85.

———, 19 April 1982: "Louisville Teletext Test Begins with 150 Home," p. 39.

———. 5 July 1982: "CBS, NBC Set to Launch National Teletext Service," p. 12.

———. 23 August 1982: "Keyfax Set for Launch in November," p. 4.

———. 11 October 1982: "Time Plans Launch of Teletext Trial," pp. 1 & 12.

———. 9 September 1983: "Zenith to Have Advanced Teletext Decoder in 1984," p. 4.

———. 10 January 1983: "Taft Zenith Announce Teletext Hardware Deal," p. 1.

————. 14 February 1983: "CBS To Launch Teletext in April," p. 36.

————. 18 June 1984: "Satellite-Delivered Text Service Signs 4 Carriers," p. 18.

————. 3 September 1984: "Teletext to Threaten Papers, Newscasts, IRD Study Predicts," p. 59.

————. 17 September 1984: "Teletext Service Said to Reach 50 Percent of U.S.," p. 14.

"Nation's First Local and Network Extravision Teletext Service Inaugurated April 4 by WBTV, Charlotte, and CBS Television Network." *CBS Press Release*, 4 April 1984.

"NBC Launches Its Network Service," *Press Release*, 1983.

"NBC Pact Sets Stage to Launch a National Teletext Magazine." *Electronic Engineering Times* (February 1982): p. 56.

"NBC Teletext—A History," *Press Release*, 1983.

"NBC Teletext Currently Offering," *Press Release*, 1983.

"NBC Teletext Unveils First Paid Advertisement at NAB," *Press Release*, 28 April 1984.

"NBC's Barbara Watson," *Press Release*, 1983.

Norris, Eileen. "Teletext Goes Local." *Electronic Media* (April 1986): pp. 18–22.

*North American Teletext: Reaching New Markets.* Ottawa: Canadian Department of External Affairs, 1982, pp. 1–33.

Ohlin, T. "Videotex and Teletext in Sweden—A Nation Decides." *Viewdata '81 Proceedings.* Online Conferences, Ltd., London, October 1981, pp. 215–30.

"Online." *Broadcasting* (30 July 1984): p. 81.

O'Reilly, Bob and Marius Morais. "Broadcast Telidon in Canada." *Canadian Broadcasting Corporation,* 1981, pp. 1–4.

Parr, Jan Margaret. "Can Teletext Hit Home?" *SAT* (April 1984): pp. 8–10.

Percelay, David. "Strategic Planning for a Major Market Trial of Broadcast Teletext." *Videotex '81 Proceedings, Toronto.* Online Conferences, Ltd., London, UK, 20–22 May 1981, pp. 71–76.

Pergament, Alan. "Channel 4 Starts Electronic Newspaper." *Buffalo News*, 2 August 1984, p. B4.

Pfister, Larry T. "Teletext . . . Its Time Has Come." *National Cable Television Association Conference, Las Vegas.* 2 May 1982, pp. 1–8.

Pollack, Andrew. "Teletext is Ready for Debut." *New York Times*, 18 February 1983, p. D6.

Raggett, Michael R. "Teletext's Uncertain Future." *Computer Graphics World* (February 1985): pp. 73–74.

"Random Thoughts." *Broadcast Week* (14 May 1984): p. 56.

"Results of Los Angeles Teletext Research Project." *CBS Press Release*, September 1982, pp. 1–3.

Roizen, Joseph. "Teletext—A Service That's Coming of Age." *Educational and Industrial TV* (September 1982): pp. 39–40.

"Role Seen For 'Teletext' In Home Banking." *EFT Report* (January 1983): p. 3.

Shaw, Sidney. "Teletext Service Will Be Available by Next Year." *Videoprint* (3 May 1984): pp. 11–12.

Spillman, Susan. "Time, Matsushita Link on Teletext Terminals." *Electronic Media* (16 December 1982): p. 20.

Storey, J. R., A. Vincent and R. Fitzgerald. "A Description of the Broadcast Telidon

System." Ottawa: Federal Department of Communications, 18 August 1980, pp. 1–6.

Storey, J. R., H. G. Brown, C. D. O'Brien and W. Sawchuk. "An Overview of Broadcast Teletext Systems for NTSC Television Standards. Ottawa: Department of Communications, February 1980, pp. 1–21.

"Teletext and the Broadcast Revolution." *North American Teletext: Reaching New Markets.* Ottawa: Department of External Affairs, 1981.

"Teletext and Prestel—User Reactions—Report One—Teletext Users." *Communications Studies and Planning Ltd.,* London, February 1981, pp. 1–17.

"Teletext Experimentation in the U.S." *Electronic Publishing in the Home of the 1980's.* Donaldson, Lufkin & Jennrette, 1981, pp. 1–4.

"Teletext User Survey—Ceefax and Oracle User Reactions." Washington: Videotex Industry Association, 1982, pp. 1–23.

"Telidon Chosen for First U.S. Consumer Trial of Teletext." Canadian Department of Communications. 6 June 1980, news release, pp. 1–2.

"Telidon in TV Ontario." *SMPTE Journal.* May 1982, p. 469.

*Teledon Reports.* Ottawa: Department of Communications.
—No. 2, June 1980, p. 1.
—No. 5, May 1981, pp. 1–2.
—No. 7, August 1981, p. 15.
—No. 9, June 1982, p. 11.
—No. 10, March 1983, pp. 5–6.
—No. 11, March 1984, pp. 9–11.
—No. 12, October 1984, p. 8.

Thomas, William L. *Full Field Tiered Addressable Teletext.* Glenview, Ill.: Zenith Radio Corporation, 1982.

Timasheff, Nicholas. *Sociological Theory: Its Nature and Growth.* 3d ed. New York: Random House, 1967.

"Time, Inc. Adopts North American Broadcast Teletext Standard for Its National Service." *Time Incorporated News Release,* 20 November 1981, pp. 1–2.

Tomita, Tetsuro. "Japan: The Search for a Personal Information Medium." *Intermedia* 7, no. 3 (1979): pp. 10–18.

Tracewell, Nancy. "There's Plenty of News in Those Black TV Lines." *Business First of Buffalo* (19 November 1984): p. 10.

"Two Nets Taking Leap into Teletext." *Electronic Media* (8 July 1982): p. 18.

Unger, Arthur. "Big Broadcasters Join Up to Push Their Version of Teletext TV News Service." *Christian Science Monitor,* 5 April 1984, pp. 20–21.

*U.S. News and World Report* (28 January 1985): "As The Video Craze Captures U.S. Families," pp. 58–59.

Vermilyea, David. "Where Will Teletext Be in 2000 A.D.?" *Educational and Industrial TV* (July 1980): p. 23.

*Videoprint.* 10 January 1983: "Status of the Videotex Industry, p. 1 +.

——. 9 January 1984: "Charlotte to Get NABTS Teletext," pp. 13–15.

——. 9 January 1984: "NABTS Terminals to Appear, More on the Way," p. 1.

——. 8 June 1984: "CBS Pushes Extravision," p. 40.

——. 8 October 1985: "Tampa Hosts a Teletext Startup," pp. 4–5.

*Videotex Viewpoint.* April/May 1985: "Broadcast Teletext: Implications of a Scarce Resource," chapter 3.

————. April/May 1985: "Business Teletext," pp. 33–34.

"Viewer Usage Study." *Electronic Publisher* (16 January 1984): p. 3.

"VSA to Promote NABTS; Announces NBC Contract." *Broadcast Management Engineering* (March 1983): pp. 35–36.

Weiss, Henry. "Teletext Viewed as First Artillery in Videotex Revolution." *Management Systems Information Week* (21 April 1982).

Weiss, Merrill and Ronald Lorentzen. "How Teletext Can Deliver More Service and Profits." *Broadcast Communications* (August 1982): pp. 54–60.

"Who Will Dominate the Home of the Future Market?" *Cable Communications* (August 1982): pp. 18–19.

Wise, Deborah. "Time Teletext to Use Panasonic Terminals." *Infoworld* (November 1982): pp. 22–24.

Wolf, Mark. "CBS Station Debuts Local Teletext Service." *Electronic Media* (12 April 1984): pp. 24–25.

Zacks, Richard. "Teletext Service Means Business." *Multichannel News* (20 November 1982): p. 9.

# Index

181